Fragrance Sense

Fragrance Sense

by Theresa Davis with Wendy Leigh

FAWCETT COLUMBINE • NEW YORK

A Fawcett Columbine Book
Published by Ballantine Books
Copyright © 1985 by Theresa Davis

Library of Congress Catalog Card Number: 84-91678
ISBN: 0-449-90107-6

Text design by Michaelis/Carpelis Design Associates
Cover design by Georgia Morrissey
Manufactured in the United States of America
First Edition: June 1985
10 9 8 7 6 5 4 3 2 1

Contents

Acknowledgments

Writing a book involves many special people. It is with deep appreciation that I acknowledge those who gave so generously of themselves.

For their friendship, faithful assistance, and valuable input: Melody Parrott, Chuck Carpenter, Mireille Noone, and Virginie Ferry.

To Mel Berger of the William Morris office; my editor, Leona Nevler; and my attornies, Jan Harris and Irving Kaler, whose beliefs, guidance, and hard work were much valued and appreciated.

A special thanks to Cher and my mother, Peggy, for their unending support and continued encouragement from start to finish.

And, lastly, to the fragrance companies, perfumers, and the Fragrance Foundation, especially its director, Annette Green, who have shared their time, knowledge, and enthusiasm for this book with me.

INTRODUCTION

Perfume is bottled magic. Webster's dictionary more prosaically defines "perfume" in two ways as: 1) A pleasing smell or odor; sweet scent, as of flowers; fragrance; or 2) A substance producing a pleasing odor; especially a liquid extract of the scent of flowers or a substance like this prepared synthetically.

The word "perfume" derives from Latin words meaning "through smoke" and dates back centuries to the Middle East when the custom was for people to burn fragrant resin and wood in the hope of persuading the gods to grant them happiness, health, and long life.

I am a passionate believer in the powers of perfume. When I wear it I feel feminine, seductive, even beautiful. I think perfume makes a woman feel better about herself, gives her confidence, and brings back memories of the best times of her life. It also glamorizes her image, enhances her attractiveness, and accentuates her femininity. Perfume is no longer a luxury. I believe every woman should wear perfume—unless she is allergic to it.

I first discovered perfume when I was seven years old. Just before a party, my mother led me into her bedroom and sprayed me with Chanel No. 5. I shrank back a bit and didn't understand why I had been spritzed with a cold, wet mist. But the moment I smelled the perfume, I felt *fantastic*. That delicious smell was actually me! I was instantly transformed. I felt grown-up, mysterious, and totally feminine. And that day, my passion for perfume was born.

Since then, I've studied perfume, lectured about it, and founded my own perfume business. Along the way, I learned that the majority of women are totally confused about just how to use perfume to their best advantage. So I became a perfume counselor and advised hundreds of women on how to buy, apply, choose, and vary their personal fragrances. I want to share that advice with you and enhance your knowledge of woman's most valuable accessory—perfume.

Maybe you never thought of perfume as an accessory, but, right along with clothes, jewelry, and makeup, it can prove to be a very valuable addition indeed. We have all been educated about the best colors to wear, the most flattering hairstyles, and the type of clothes to choose for our body shapes. I am going to show you how to enhance yourself further by developing your own fragrance sense.

How do you know if you need fragrance sense? Answer the following questions. If you see yourself, read on!

1. Can a salesperson push you into buying a fragrance?
2. Are you receiving value for your fragrance purchases?
3. Are several fragrance mistakes sitting on your dresser?

4. Are you constantly re-applying your fragrance?
5. Have you always worn the same fragrance?
6. Do you purchase a fragrance because of its pretty bottle?
7. Are you "known" by a particular fragrance or is a fragrance known among friends as "yours"?
8. Are you always admiring someone else's fragrance?
9. Are you constantly buying bottle after bottle of fragrance?
10. Do you wear a fragrance because it was given to you?

If, like most of my clients, you answered yes to several of these questions, you need to develop your own personal fragrance sense.

Everyone makes fragrance mistakes, and I, for one, have made plenty. As a very young teenager I would wear Estée Lauder's Youth Dew and Bal à Versailles by Jean Desprez—both fine, classic fragrances. The trouble was that I drenched myself in these heavy scents, believing that I would smell more like a real woman. At fifteen, I received my first perfume from a boy—an Eiffel Tower-shaped bottle, from Paris, which he bought at the World's Fair in San Antonio, Texas. The scent was atrocious. I wore it though, everywhere, just to please him. And even I have been known to give a gift of perfume simply because the box was pretty. I am as attracted to pale pink and light blue boxes as a moth is to a flame. I never once—for more than a minute—even considered the scent itself.

The perfume choices I mentioned were made a long time ago—before I knew better. I still wear Estée Lauder's Youth Dew, but in moderation and for appropriate occasions. You can avoid falling into similar traps by taking four steps to developing your fragrance sense.

Step One is learning how to save your perfume dollar. Step Two is learning how to choose the right fragrance strength. Step Three is selecting your own fragrance. Step Four is designing your own *personal* fragrance wardrobe so that it will reflect your own individual needs, personality, and lifestyle.

After you have followed my four major steps toward developing your fragrance sense, you will know what it takes to make an impact with perfume. You will know exactly how to perfume yourself to success. Remember —it's a great big billion-dollar perfume world out there, ready and eager to confuse everyone. My aim is to clear up some of that confusion.

Before going on to my four steps, I want to discuss the hard facts that often are hidden under the mystique and linguistic acrobatics of the perfume advertisements. After that, we'll get more personal....

PART I

About Perfume

CHAPTER · 1

The History Of Perfume

Perfume is practically as old as time itself. In Egypt's First Dynasty, about 5,000 years ago, it was the custom for the nobility to have fragrances buried with them. When archaeologists opened King Tutankhamen's tomb, they found airtight vases of fragrant aromatics around the Pharaoh's remains. We know that King Tut was buried around 1350 BC, which conclusively confirms that man has been using the powers of perfume for millennia.

One of the most famous ancient Egyptians to use perfume was that siren of the Nile, Cleopatra (who, by the way, wasn't at all beautiful). Cleopatra was sensible enough not to wait until she was dead before using perfume to her best advantage, however. Julius Caesar and Marc Antony were both seduced in Cleopatra's bedroom, where she kept a supposedly fatal blend of fragrances. Cleopatra was convinced that the scent created a powerful allure for men, and judging by history, she doesn't seem to have been wrong.

Egyptian *men* weren't above using perfume for power either. King Nefertum had himself anointed with aromatic oils. His odor was believed by his subjects to be the source of what they considered his power as a god.

Perfume wasn't just an Egyptian passion, though. In Rome, emperors used perfume all the time, and their courtiers invariably imitated them. In sixteenth-century Persia, a princess celebrated her marriage by having a canal dug and filled with water and rose petals. When the newlyweds rowed across the perfumed canal, they discovered a beautifully scented oily scum floating in the canal. That scum was the basis of oil of roses, which the Persians sold to the world for centuries afterward.

Not all of the history of perfume is glamour and romance. King Louis XIV of France used to sit on lavender-stuffed pillows because he suffered from gas. The Emperor Napoleon rejected perfume altogether, writing to his wife Josephine, "I'm coming home—don't wash for three days," obviously preferring a more natural scent.

Other historical notables were more delicate. George Washington was so dedicated to being elegantly scented that he even had his own fragrance blended for him exclusively by the perfumers Caswell-Massey, who are still in business today. Elizabethan ladies loved perfume and wore small packets of fragrant flowers in their cleavages. Victorian ladies used vanilla essence as a way of smelling seductive.

Royalty has always luxuriated in perfume. One Queen of Hungary added the essence of rosemary to her bath water. Hungary Water, as it was then called, is the oldest known perfume water and is still being manufactured today. English royalty also loves perfume. In fact, it is a British custom to pelt royalty with flower petals during a coronation or wedding. When Prince Charles married Lady Diana Spencer in 1981, the royal couple's attendants (children, who were so adorable that they almost upstaged the bride) carried baskets of rose petals and tossed them at the feet of the bride and groom.

Lady Diana also commissioned Floris, a well-known fragrance company famous for its perfume waters, to create the perfume Wedding Bouquet for the special day in her life.

The Arabians began the perfume industry by developing the method of distilling the essence of flowers (especially roses) so that the fragrance was turned into the liquid form that we know today. But perfume was finally perfected in the French town of Grasse, where the countryside is filled with roses, jasmine, and lavender. Grasse was originally known for manufacturing scented leather that was used to make gloves. As the industry expanded, the essences of many flowers were distilled and extracted, and perfume was born. Today Grasse is the world center of perfumery, which has become a multi-billion-dollar business.

CHAPTER · 2

How
Perfume
Is Made

One of the most romantic aspects of the Grasse perfume industry is the harvest, when the flowers are picked at dawn by lantern while the petals are still dripping with the night's dew. After that, the essence of the flower is extracted.

Many people think that the process stops there—that perfume is made up of crushed flowers. But that is economically impossible. It would take *two thousand* pounds of rose petals to produce only *one* pound of rose oil. This would cost $2,000 to $5,000, depending on the variety of rose. And that pound would only be the *essence* of roses—which still wouldn't make a perfume. It takes seventy-five to two hundred different essences to make a fine fragrance, so you can imagine how much that would cost if only the essence of flowers were used. The truth is that a perfumer blends essences and fixatives (more about them in a minute) together to create a fragrance. A perfumer sometimes spends many years creating a new fragrance.

Once the flowers are gathered, they are either distilled or macerated in vats so that the oils are drawn off. Delicate flowers are distilled—their petals are brought to the boiling point in water and then the desired essence is isolated as the vapor condenses. Less delicate flower petals are macerated—soaked in huge vats of suet or lard. Fresh petals are added as old ones are taken away, so that the suet or lard is eventually turned into a fragrant oil essence.

Once the perfumer has extracted the oils, fixative is added to ensure the perfume's staying power. Fixatives used to be obtained from whales, civet cats, and male musk deer, but today they are composed of often evil-smelling chemicals. After the essence and the fixative have been mixed, the perfume is subjected to the scrutiny of the perfumer referred to in the industry as "the Nose." The Nose's job is to decide whether the perfume fulfills the requirements of the manufacturer in that particular fragrance. The Nose is responsible for the perfume we buy in our department store.

The first time I visited a perfumer's laboratory, I was awed by row after row of liquid-filled amber glass bottles. Each bottle represented a different kind of smell. A Nose has approximately 3,000 raw materials with which to construct a perfume. The Nose's salary is based on the number of raw materials he or she can correctly identify by sense of smell only. The blending together of these raw materials can create endless possibilities for fragrances. The lab is almost hospitallike in appearance, with paper test strips, glass beakers, filters, funnels, and eyedroppers visible everywhere. Even the Noses wear white coats while working out their formulations.

It is important to understand that the creation of perfume is very complex. Some people have likened it to a symphony.

Most perfume is created to level off, to change character as it lives on your skin. It can adapt to each individual and blend with her chemistry without losing its own unique quality. The process that creates this effect is talked about in terms of "notes." The perfume is brought together in layers of evaporation. Perfumers record the evaporation time of the layers (or notes) on an "evaporation scale." The times vary according to the composition of the fragrance and the body chemistry of the person wearing it. The top note is the first impression we receive of a perfume. This lasts for less than six seconds. The middle note reflects the true scent of a fragrance and will remain on your skin for approximately one hour. The bottom note usually contains the fixative of the fragrance and lasts the longest of all —three to four hours on the skin.

Some perfumers are now making fragrances in a linear way so that the last layer of the perfume is constructed first. This creates instant appeal; the scent changes very little from the moment you put it on.

CHAPTER · 3

How Perfume Works

Y our response to a given scent begins in your brain. There are ten to twenty million olfactory receptor cells in a dime-sized space at the top of your nasal cavity. These cells are capable of detecting four thousand separate smells. It is this area that, like a computer, identifies and reacts to the scent you are wearing.

John Labows, Jr., who is with the Monell Chemical Senses Center (which researches smell), shows how the sense of smell responds to chemical stimuli such as perfume by explaining that: the sense of smell can provide a different sensory experience with each molecular structure. The odorant must have a high vapor pressure and proper solubility to dissolve in the mucous layer. In the mucous layer, there are cellular projections called cilia, which interact with the odorant through proteins in the cell membrane. Each odor molecule may visit more than one type of protein receptor.

This interaction initiates a series of neurological events which we recognize as odor perception.

By the way, not everyone is capable of smelling perfume. Some people have a condition called anosmia, similar to color blindness except that it applies to smell. These people have a limited sense of smell and can't differentiate between certain scents. A smell survey, conducted by the Fragrance Foundation and the Center for Molecular and Sensory Disorders, found that, at any given time, approximately 4.7 million people suffer from smell distortions. What this means to you and me is that it is much more common than we ever before realized. To have a distorted sense of smell may well be a drawback in choosing fragrances and fragrance products.

Even if you have a weak sense of smell, it is possible to increase your smelling power. Often, the smell of something strong, pungent, or sharp will reverse some of that loss. You can wake up your sense of smell every morning by sniffing a "bright" or strong fragrance. (A citrus scent is a good example.) The sense of smell peaks by early afternoon. If you start the day awakening your senses, by early afternoon they will be responding as you want them to. According to the Fragrance Foundation, a nonprofit information center, "Current research concludes that constant, creative use of all our senses can keep them at peak performance well into old age."

Most people's sense of smell peaks around the age of thirty and begins to diminish by age fifty. Women have a keener sense of smell than men because of hormones, estrogen primarily. Researchers for the Fragrance Foundation and Georgetown University have found that, associated with the secretion of estrogen, smell acuity increases during the first half of every month. A decrease in smell acuity during the second half of every month is associated with the secretion of another hormone, progesterone. But the average level of acuity for a woman is, in general, stronger than that for a man.

Dr. George Dodd of the University of Warwick, Coventry, England, says that our olfactory sensors are continually transmitting a variety of messages to the brain —recognition, for example. Humans have the ability to distinguish one person from another on the basis of odors. Odors can even trigger emotions in us.

Scientists have known for many years that scent can attract or repel the opposite sex. Research has proved that insects and mammals secrete something called a pheromone just before they mate. They believe that these pheromones are so strong that they actually are the primary reason for sexual attraction. None of the scientists, though, are a hundred percent sure whether humans also secrete pheromones. But some modern fragrances (such as Jovan's Andron) now contain a synthetic chemical pheromone that may well work to attract the opposite sex.

PART II

The Four Steps To Fragrance Sense

CHAPTER · 4

Step One: Saving Your Perfume Dollar

On the surface, shopping for perfume sounds enjoyable and uncomplicated. After a hard day's work, you decide that you deserve a treat. What better treat than perfume—sheer bottled luxury? So you drift into the perfume department of your local store and suddenly find yourself in paradise. The lighting is soft, the ambiance elegant, and you are surrounded by row after row of beautifully seductive-looking boxes of perfume. What could be simpler than buying perfume?

Think again. There are more than 600 different perfumes on the market today—and almost every one of those little bottles costs big bucks. The perfume industry grosses more than two billion dollars a year! Part of that revenue is plowed back into the industry specifically so that you will be lured into spending more money on perfume. Each year, perfume advertisements become more and more alluring. Perfume boxes are made to be more enticing and perfume bottles more luxurious. Then there are the celebrities—sultry beauties like Sophia Loren, Candice Bergen, and Joan Collins—who lend their superstar luster to promoting perfumes.

In Step One, I want to help you avoid throwing your money away on the wrong perfume for you. The perfume industry may make billions of dollars, but presumably you don't. I want to save you some of your money.

THE PERFUME PITFALLS

* The Advertising. The first perfume pitfall is buying a fragrance because you liked the way it was advertised. Not that I would blame you. Perfume advertisements are every bit as seductive as the perfume companies intend them to be. It has been said that more money goes into advertising, design, and manufacture of the bottle and box than is spent on making the perfume itself. The fact is that perfume that sells for $100 a bottle at retail costs the store about half that amount. The perfume itself (the essential oils, fixatives, and other ingredients) may have cost the manufacturer only $5. The manufacturers spend much more than that marketing perfume. So it is hardly surprising that many people fall for it and pick their perfume on the basis of the package and the advertising, not because of the scent itself. Ads cannot convey what a fragrance smells like, so they have to create an image of what the perfume will do for you. Copy featured recently in magazines includes phrases such as:

> "The fragrance of the rich and outrageous."
> "The gorgeous, sexy young fragrance."
> "Never has a fragrance provoked such emotion."
> "The perfume of eternal youth."
> "The perfume that makes one question the value of civilization."

Even though the words sound captivating, make sure they don't lead you astray.

* The Name. The second perfume pitfall is buying a fragrance because you like its name. The name of a perfume really has *nothing* to do with what it can (or can't) do for you, but names can seduce people into buying fragrances. French names have always sounded elegant or romantic. Now there are movie star names and designer names as well. Oscar de la Renta, named after the designer, has been the best-selling perfume in the United States since 1980. The names of perfumes hint at high fashion, foreign intrigue, and even diamonds now that the jewelers Van Cleef & Arpels, Cartier, and Tiffany all have their own line of perfume. Diane Von Furstenberg even included a poem with her perfume Volcan d'Amour, which she dedicated to her boyfriend. Names may sound alluring and enticing, but they don't tell you what's inside the bottle. Only your *nose*, not Sophia, Candice, Joan, Oscar, Calvin, or Halston, can decide if the fragrance is one you will like.

  A perfume shouldn't be under suspicion just *because* its name evokes glamour, either. Obviously, many women prefer Oscar de la Renta for its lovely bouquet.

* The Packaging. The third perfume pitfall is picking a perfume because you like the bottle or the box. Fragrance prices increase every year. Part of those increases can be accounted for by the intricately beautiful perfume bottles currently being manufactured.

  Market studies of packaging have indicated that the more beautiful a container, the more we want to possess it. Perfume companies are highly aware of this. The curvaceous bottle of Halston is so appealing that the Halston advertisements feature it without any accompanying copy, knowing that it will convey the necessary message of luxury and magic. Molinard de Molinard unveiled a limited edition perfume in a crystal bottle cast from a 1929 Lalique mold. Sculptor Nikki de Saint Phalle created a fragrance bottle complete with intertwining serpent for the Cartier fragrance that bears her name. Le Must de Cartier scent is available in a removable glass bottle encased in an eighteen-carat gold frame. Bijan, the men's clothing store, has actually issued a fragrance that is priced around $1,500, comes in a crystal bottle, and is insured by Lloyd's of London.

Succumbing to the perfume bottle pitfall is very common. When Jean Patou, who make Joy, advertised a limited edition of their perfume in a Baccarat crystal bottle, sales nearly doubled.

There are beautiful perfume bottles, and there isn't anything wrong in buying them, as long as you know that *that* is what you are doing. Perfume bottles do make attractive ornaments, but the bottle's attractiveness has nothing whatsoever to do with the fragrance itself. So avoid perfume pitfall number three —buying a perfume merely because you like the bottle.

* <u>The Price</u>. Perfume pitfall number four goes back to the Joy story. Patou advertises Joy as the most expensive perfume in the world. When the Baccarat bottle increased this already substantial price, sales nevertheless doubled. This does suggest that the more expensive a perfume, the better it sells. In fact, I often wonder how many people would buy the perfumes they buy if those perfumes came in plain bottles and cost $7 instead of $70. Remember that no matter how expensive a perfume is, it still may not be right for you.

* <u>The Salesperson</u>. The fifth perfume pitfall concerns fragrance salespeople. First, I want to make it clear that the salesperson can be your most helpful resource in choosing a perfume. She is an expert, having undertaken at least two to three training periods from her company each year. But that's the problem —inside *her* company. Ever notice the perfume counters in department stores? Side by side, up and across

the aisles, each fragrance has its own section of counter *and* its own salespeople, too. The salesperson knows what she is selling, and she is not interested in selling the fragrance sitting on the next counter. If you want an oriental fragrance (page 33) and you don't like Company X's oriental, the salesperson might show you a floral fragrance from Company X, but she won't mention Company Y's oriental sitting on the next counter. And you might miss buying what could have been your favorite perfume. Resist the hard sell; if it is not what you want, learn to say no. I used to have a dresser full of perfumes that I didn't really like but had bought because I was told, "They smell wonderful on you," or "This is *the* latest. Everyone is wearing it." Trust your own judgment, and your own nose.

* <u>The Other Person's Taste</u>. The sixth perfume pitfall is choosing a fragrance because it smells good on someone else or because someone else (a boyfriend, maybe) likes the way it smells on you. Fragrances smell different depending on who is wearing them. Your body chemistry (again) plays an important role in determining what a perfume will smell like on you. I advise all my clients to sample different perfumes.

I also advise women to choose a fragrance *they* like instead of their mother's, sister's, or best friend's choice! A boyfriend or husband can be diverted to a fragrance that pleases you both.

So, forget about the advertising, the name, the packaging, the price, the salesperson, your best friend, and your boyfriend. Concentrate on the fragrance itself —and on *you*. The emphasis on perfume should always be what *you* want. Focus on the scent and not the packaging—because that is the first step to saving your perfume dollars.

RULES FOR BUYING PERFUME

I've told you what *not* to do; now I want to suggest what I think is the best way to buy perfume. At this point, I am not going to advise you on what particular scents to buy; that comes in Steps Three and Four. First I want to teach you the basic principles of buying *any* perfume; after that we'll go into more detail about what fragrance is best for you.

Picture this situation: you go into the perfume department, clutching your $20 to $40 in your hand, ready to buy a new fragrance. Trouble is, you have *no idea* which perfume to buy. You are determined not to be influenced by the various pitfalls I pointed out, but you have no guideline to fragrance other than your own nose. And I would bet a thousand dollars that you don't really know how to use it!

Smelling perfume isn't as simple as it sounds. Women do have a keener sense of smell than men, and they exercise it more often than men do. But even at its zenith, very few people really do have a good sense of smell, and most people's sense of smell declines after they have sniffed three different scents. That is a very important fact to remember.

* Sample only three scents at a time. There is absolutely no point whatsoever in going to a perfume counter planning to try every fragrance in sight because after the third scent, everything will be severely distorted. Your sense of smell will be dull and confused. Never smell more than three scents at a time because after that you won't get the *true* smell of the fragrance. Instead, go back to the store the next day and try another three scents.

* Always try on a fragrance. Never sniff test strips or smell directly from the bottle. The best way to try out the fragrances is to ask for a sample and take it home to experiment. The second choice would be to apply the products to your skin while you are in the store. Apply one fragrance to the right wrist, one

fragrance to the left wrist, and one to the inside of your elbow. Remember which fragrance you put in each place so that you will be able to judge which one you like the best.

* <u>Never smell a fragrance in the first six seconds after applying it</u>. Remember, the alcohol's "bite" is evaporating, and after fifteen to twenty minutes the character of the fragrance will start to change.

* <u>Wait one to two hours</u>. I always sample first, shop or have lunch next, then go back to the perfume counter to make my fragrance purchase. By waiting at least one to two hours, you will have a clear idea of what the fragrance will smell like on you, and you will better understand the complexity of the scent.

Other tips on saving your perfume dollars:

* <u>Buy larger-sized bottles</u>. Do this if you can manage to because you will save at least one-third on double the amount of fragrance.

* <u>Look for perfume specials in your department store</u>. All fragrance companies usually have special promotions throughout the year—but January is the best time for fragrance bargains.

* <u>Remember to buy refills</u>. They can save you at least one-third off the total price of replacing the perfume bottle. After all, why spend money on a pretty bottle when you have another one just like it at home?

CHAPTER · 5

Step Two: Choosing The Right Fragrance Strength

"**H**elp," said Cheryl, a striking-looking woman in her mid-twenties, "perfumes don't seem to *last* on me. What's wrong?"

This has been the number-one complaint women have brought to me whenever I give a lecture or personal fragrance consultation. Fragrance genuinely *does* wear off very quickly on some people. These women may start the day drenched in perfume but find that, in an hour or two, the smell has completely disappeared. This happens because individual body chemistry may actually cause fragrance to evaporate very fast. The technical term for all this is someone who "throws off" perfume.

In Chapter 2, we talked about the perfumer's evaporation scale for fragrance. For most women, evaporation times for the perfume layers are:

0–6 seconds: Alcohol's "bite" evaporates.
After 15–20 minutes: Top layer starts to change.

After 1 hour: Middle layer begins to change.
After 3-4 hours: Bottom layer fades.

For a woman who throws off her scent, the evaporation rate almost quadruples:

0–6 seconds: Alcohol's "bite" evaporates.
After 5 minutes: Top layer starts to change.
After 15 minutes: Middle layer begins to change.
After 30 minutes–1 hour: Bottom layer fades.

YOUR BODY CHEMISTRY

Let's examine a few of the reasons women might throw off scent. I mentioned body chemistry as the main culprit, but what makes your body chemistry produce the chemicals that fight off perfume? One of the major causes is acid. Almost everything we do produces acid within our bodies, and it comes through to our skin. Lemony-tasting skin is a sure sign of acid build-up.

Medication also works overtime in our bodies to help throw off perfume. Birth control, anti-allergy pills, vitamins, aspirin, even high blood pressure medication can affect your body's chemistry. Diet and stress play important roles, too. Stress is no longer just a man's illness. After talking with Cheryl for a few minutes, I realized she had fallen into the eighties dilemma of trying to do it all—become a superwoman. The mother of an active four-year-old child and the wife of a promising young attorney, she was working in a high-pressure job that required her to produce constantly. Nursery school, cocktail parties, clients calling at home after hours, grocery buying, cooking and dinner planning, and the pressure from her job amounted to stress with a capital S.

Eating spicy foods or foods high in fat causes fragrances to evaporate more rapidly, too. Eating them on a regular basis almost guarantees your skin's throw-off propensity. We can also add dry skin and pregnancy to our list of reasons a perfume has no staying power on a particular person. I'm not going to try and play doctor—I can't tell you how to avoid acid build-up—but I can show you how to compensate for it.

THE SCIENTIFIC PERFUME TEST

Don't mistake a weak sense of smell for a throw-off body chemistry type. I have designed the Scientific Perfume Test based on information gathered from hundreds of women who throw off their perfumes. This easy, short, multiple-choice questionnaire will reveal if *you* actually throw off your fragrances.

The Scientific Perfume Test will help distinguish between women who throw off fragrance and women without a keen sense of smell. The fact that you might be unable to smell the fragrance you wear doesn't mean that other people around you can't.

Choose the *one* answer that comes the closest to describing you best.

1. Is your hair color
    a) Red   b) Blonde   c) Brown or brunette
2. Is your skin type
    a) Dry   b) Normal   c) Oily
3. Do you have acidic skin? Find out by licking the inside of your wrist. Does it taste
    a) Lemony   b) Salty
4. Do you take hormonal medications? For example, birth control pills?
    a) Yes   b) No
5. Do you take other medications on a regular basis?
    a) Yes   b) No
6. Is the tone of your skin
    a) Fair   b) Medium   c) Dark
7. Is your diet high in fruits?
    a) Often   b) Occasionally   c) Never
8. Is your diet high in fats?
    a) Often   b) Occasionally   c) Never

9. Is your diet high in spices?
   a) Often   b) Occasionally   c) Never
10. Do you perspire?
    a) Often   b) Occasionally   c) Never
11. Is your age group
    a) Teens to 30   b) 30 to 45   c) 45 plus
12. Do you smoke?
    a) Yes   b) No
13. Do you drink alcoholic beverages?
    a) Yes   b) No
14. How often do you drink alcoholic beverages?
    a) At least once a day   b) 3–4 times a week
    c) Seldom–twice a month   d) Never
15. Do you consider yourself easily excitable or highly strung?
    a) Yes   b) Sometimes   c) No

Give yourself points for the answers as follows:

1. a–three points; b–two points; c–one point
2. a–three points; b–two points; c–one point
3. a–three points; b–two points
4. a–three points; b–one point
5. a–three points; b–one point
6. a–three points; b–two points; c–one point
7. a–three points; b–two points; c–one point
8. a–three points; b–two points; c–one point
9. a–three points; b–two points; c–one point
10. a–three points; b–two points; c–one point
11. a–three points; b–two points; c–one point
12. a–three points; b–one point
13. a–three points; b–one point
14. a–three points; b–two points; c–one point; d–one point
15. a–three points; b–two points; c–one point

Scoring

15 to 25 points: This is the normal range, even for those without a keen sense of smell.

25 to 35 points: Occasionally you throw off scent.
35 to 65 points: Definitely a throw-off.

If you scored from one to nineteen points, your skin holds a fragrance easily. You can use heavy and light fragrances with equal success. Follow the Perfume Application Guide on page 53.

If you scored twenty or more, your skin holds a fragrance for relatively short periods of time. Follow the Throw-Off Application Guide on page 54. When you get to the Fragrance Guides later in the book, try and stick to fragrances marked H. You may think that this limits you too much, but if you use perfumes not graded H you will have to apply a great deal for the fragrance to last, which will end up being very expensive.

PRODUCT GUIDE

If you "throw-off" scent, you should purchase fragrances by strongest concentrations. By concentration, I mean the ratio of actual perfume essence to the alcohol (spirit) and water content. Most fragrances—except oils—contain some alcohol in order to give a "lift" to the scent.

It is often confusing to enter a store and select a fragrance with a strong concentration. Which kind of product becomes the best choice—perfume, extrait, essence mist, eau fraiche, or esprit? Do you buy a splash, a veil, a gel, a touch-on, or a natural spray? The Product Guide will steer you through all the different fragrance products, with brief descriptions and definitions. I have started with perfume and worked down to bath beads, so the fragrance decreases in strength the farther down the list you go. The list is a quick and easy way of telling you what exactly all the perfume grades mean. You can see at a glance which products are more concentrated and how broad a choice of fragrance you have.

There are three main families of fragrance: Perfume, Eau de Toilette, and Eau de Cologne. All products are listed in order of strength, the most concentrated (strongest) first.

Perfume

Parfum/Perfume
Perfume Concentrate
Extrait/Extract
Eau de Parfum/Perfume
Pure Fragrance
Esprit de Parfum/Perfume
Soft Perfume
Eau de Parfum/Perfume Toilette

Fragrances from the Perfume family are the most concentrated form of fragrance available. The small ½ to 1 ounce bottles of parfum concentrate represent the ultimate in a fragrance and can cost anywhere from $55 to $250. Perfumes contain 15 to 20 percent pure essence and 80 to 85 percent alcohol (spirit). Occasionally a minute amount of water is added. This strength of fragrance has the longest staying power on the skin —four to six hours depending on your body chemistry. Perfumes are available in splash, solid (usually pretty compacts), purse spray, and touch-tip (a felt-tip applicator wand).

Eau de Toilette

Eau de Toilette Concentrate
Eau de Toilette
Essence Mist
Esprit de Toilette

Eau de Toilette contains 5 to 10 percent pure essence and mostly water with some spirit added. This is the best-selling fragrance product on the market because it is medium priced. Eau de Toilettes are available in splash, spray, touch-on applicator wands, mist, and natural spray. Natural spray uses no gas to propel the liquid—think of a pump.

Eau de Cologne

Eau de Cologne Concentrate
Super Cologne
Cologne
Eau de Cologne
Eau Fraîche/Fresh
Body Fragrance
Body Freshener
After Bath Splash
Friction

Mostly water, Eau de Cologne contains 2 to 6 percent pure essence. The water content of Eau de Cologne speeds its evaporation process. Meant to be used lavishly as a refresher, this product is the least expensive. It will last one to two hours on the skin. Eau de Cologne is available in splash, spray, solid stick, natural spray, and touch-on applicator wands.

Other Fragrance Products

Bath Oil
Bath Oil Spray
Bath Silk
Body Oil

Oils contain the highest concentration of fragrance of all fragrance products other than perfumes. Bath silks glide on after bathing or showering. They contain a high concentration of oil. Each of these products is long-lasting.

Milk Bath
Crystals
Foam Bath
Bubble Bath
Sea Salts
Foaming Bath Powder

The use of these products in the bath water can leave your skin incredibly soft and sweet smelling. Almost all bath products are available in the scent of your choice.

Body Creme
Creme Oil
Hand and Body Lotion
Scented Moisturizer
Body Silk
Soft Body Veil

Cremes are more emollient than lotions, moisturizers, and veils. They also contain more concentration of fragrance than these other products.

Poudre Parfumée
Dusting Powder
Talc

Powders help scented lotions cling to the body longer, thereby leaving a faint hint of fragrance on your skin. A growing number of gynecologists would like to see powders removed from the market. They believe that the tiny particles in powder are ingested into your body through its natural openings. They also believe these particles can cause serious health problems. I advise my clients to ask their gynecologists about this effect and make their own decision regarding the use of a powder. There are now cornstarch-type powders available in some fragrances, and these are not controversial.

Body Gel
Body Shampoo
Fragrant Soap
Body Scrub
Scented Deodorant

Many fragrance companies are adding these kinds of products to their lines. They are available in your favorite fragrances.

PROBLEM-SOLVING
USING THE PRODUCT GUIDE

In applying perfumes, throw-offs should layer the different products for a long-lasting fragrance. Start with a bath oil or use a body gel in the shower. Some companies have a bath oil spray to use after bathing. After-bath splash can be used in place of a bath oil spray. Scented moisturizers, body lotions, cremes, or silks provide a faint lingering of scent all day.

Apply a strong concentration (check the Product Guide) of fragrance to *all* pulse points—to your wrists, at your bosom, the bend of your elbows, behind your ears, the nape of your neck, and the back of your knees. Heat generated at these points intensifies fragrance. Finish off with a light misting of a strong concentration of fragrance. Remember, fragrance rises. Applied just behind the ears, your perfume will rise and disappear. Applied all over, fragrance takes longer to fade.

Another way to help your fragrance last longer is to scent your undergarments. Pantyhose and lingerie can be lightly misted. I wave my garments through a line of spray mist. Be careful! Direct spraying can stain clothing. Sprinkle perfumed talc under drawer liner papers, unwrap scented soap and put it in your bureau, place empty perfume bottles in lingerie drawers. You can also hang sachet bags on hangers or splash a few drops of cologne into the rinse water of lingerie.

Never spray fragrance on your hair, though, unless it is freshly shampooed and towel dried. Oils and dirt that accumulate on the scalp can intensify and distort fragrance.

CHAPTER · 6

Step Three: Selecting Your Own Fragrance

My mother wore Chanel No. 5 all her life, but, much as I love it, I *refuse* to do the same. There are so many perfumes available on the market that I want to take advantage of this wealth of choice. I think the idea that a woman should wear one perfume—and only one perfume—all the time went out with the hoop skirt.

Fragrances fall into eight basic categories. Once you have decided which category you like best, there are many and varied alternatives in that category. In Step Three I am going to help you pick your perfume category and make you aware of all the possibilities inherent in that choice.

First, let's become familiar with the eight fragrance categories. Then I will show you how to discover which is yours.

THE EIGHT BASIC FRAGRANCE CATEGORIES

The lighter scents are generally found in the first five categories.

* Green. This is the newest, therefore youngest, of the basic fragrance categories. It was developed between 1950 and 1970 from a modern citrus accord. This category is fresh, brisk, clean, and sparkling. A few green smells you might be familiar with are lemon, lime, hyacinth, orange, bergamot, and galbanum—the smell of fresh green vegetables, like the green pea. Green notes are found primarily in the top layer of a perfume.

* Fruity Fresh Floral. Fruity fresh notes lend originality to a fragrance. Perhaps that is a clue as to why this category has been referred to as the most novel of all. Pineapple, cassis (blackberry), apple, peach, and strawberry might be used in creating a fruity perfume. "Fresh florals" refers to lily-of-the-valley and other light, springlike flowers.

* Floral. Floral is the most popular basic fragrance category. It can be a combination of several flowers or a single flower itself. It is found in the middle layer of a perfume, but it can influence the top and bottom layers as well. Jasmine, rose, lilac, violet, carnation, jonquil, muguet, ylang-ylang, frangipani, tuberose, honeysuckle, stephanotis, mimosa, and freesia are but a few of the flowery smells found in floral fragrances.

* Aldehydic Modern Blend. Aldehydes are pure aroma chemicals with no natural-based scent. They are used to give a fresh smell throughout the top, middle, and bottom layers of a fragrance creation. Some aldehydes provide the clean-laundry and just-washed-sheets smell we find in our detergents and fabric softeners. These aldehydes are usually blended with spicy scents—cinnamon, cloves, carnation, and nutmeg—to create a light, crisp, spicy perfume.

Most classic fragrances (Chanel No. 5, for instance) are found in this basic fragrance category.

* Aldehydic Woody Floral. This category combines the fresh aldehydes with oakmoss, tree moss, sandalwood, vertiver, cedar, and patchouli scents. When you think of "powdery," "velvety" perfumes, you are thinking of those in the aldehydic woody floral category.

The truly heavy and dramatic fragrances are found in the remaining three basic fragrance categories.

* Chypre. Chypre is an ancient and classic accord. Bergamot (citrus with a light note of mandarin) is mixed with oakmoss to form a contrast. (Oakmoss is a smell similar to tree moss only more refined.) The finished fragrance is reminiscent of new-mown hay and lavender bouquets. There are almost as many men's chypre fragrances on the market today as there are women's.

* Oriental. Oriental fragrances characteristically have a marked sweetness about them. Oriental is a category created from oriental odors—tonka beans (a vanilla scent), benzoin (a resin from trees that grow in Sumatra and Thailand), opoponax (another sweet resin), and ground pepper (found in the top note of an oriental fragrance). Oriental notes are almost always used in the bottom layer of a perfumer's construction of a fragrance.

* Animalic. This is the category from which perfumers find their fixatives for holding or setting a fragrance together. Musk (from the musk deer), civet (from the civet cat), and castoreum (the gland secretion from the beaver that gives him a water-repellent film for his fur) are common animalic scents. Vanilla bean, ambergris (from the sperm whale), and labdanum (a sweet, resinous incense smell) are common amber fragrances. Animalic/amber notes are also found in

the bottom layer of a perfume because they have the longest-lasting fragrance properties of all the categories. Animalic fragrances are easily recognized because of their musky odor. Musk can be added in larger quantities to fragrances from other categories to create longer-lasting smells.

Now that you are familiar with the eight basic fragrance categories, let's advance on to the Perfume Trail to discover in which category you belong! Where will the Perfume Trail lead you? To your own basic fragrance category—and *that* leads to a personal fragrance style!

* The first step is to *select three fragrances* you like and have consistently worn during your lifetime.
* Then, simply *find those three fragrances* in the Perfume Trail—a listing, in alphabetical order, of 100 popular name-brand fragrances. Beside each perfume, its Basic Fragrance Category is revealed.
* Then, *observe which category they fall into*. Chances are, at least two of the three fragrances will be in the same category.

Imagine being able to go into the store and tell the salesperson exactly what you are looking for! No more buying a fragrance because your best friend wears it, or because the bottle is pretty, or because it is expensive and expensive must mean good. The Basic Fragrance Category will become *your* base. Later on in this book, you will build your personal fragrance wardrobe on it.

Some women may find their three perfume categories do not match up in the Perfume Trail. If that happens to you, select the *one* perfume that you simply adore—not your husband, or boyfriend's favorite, and not your mother's or best friend's selection—but the one perfume *YOU* cannot seem to do without. Then follow the Perfume Trail and discover your Basic Fragrance Category.

Remember, too, there is no hard and fast rule that says you must limit yourself to one basic fragrance category. I like the oriental category best, but I also happen to enjoy wearing perfumes from the green category. When I play tennis or have lunch at an outdoor cafe, I choose green fragrances. When I dress for a romantic evening or an elegant dinner at a favorite restaurant, I wear an oriental perfume. In Chapter 7, we will discuss the different occasions you encounter in everyday living and how to choose the appropriate fragrances for each of them.

In case your favorite perfume is not among the 100 fragrances in the Perfume Trail, you will probably find it in the Name-Brand Fragrance Guide that follows it. More than 300 perfumes are listed in the guide by their basic fragrance category. But for now, let's move on to the Perfume Trail.

THE PERFUME TRAIL

Fragrances are listed alphabetically, first by product name, then by manufacturer. Basic fragrance categories are shown to the immediate right of the product/manufacturer. (Men's fragrances are marked with an M.)

Fragrance	Category
Albert Nipon by Nipon	Aldehydic Modern Blend
Aliage by Estée Lauder	Green
Ambermist by Avon	Oriental
Ambre d'Or by Parfums Poêt-À-Porter	Aldehydic Woody Floral
Amouage by Asprey	Aldehydic Woody Floral
Anais Anais by Cacharel	Fruity Fresh Floral
Angel Fire by Mary Kay Cosmetics	Aldehydic Woody Floral
Anne Klein Parfum by Parlux	Floral
Armani by Giorgio Armani M	Fruity Fresh Floral
Aromatics Elixir by Clinique	Oriental

Arpège by Lanvin	Aldehydic Modern Blend
Avenir by Mary Kay Cosmetics	Floral
Aviance by Prince Matchabelli	Aldehydic Woody Floral
Bal à Versailles by Jean Desprez	Oriental
Bijan by Bijan M	Aldehydic Woody Floral
Bill Blass by Revlon	Floral
Black Sapphire by Revlon Colourscents	Oriental
Blue Grass by Elizabeth Arden	Floral
Cabochard by Grès	Chypre
Cachet by Prince Matchabelli	Chypre
Calais by Mary Kay Cosmetics	Floral
Calvin Klein by Klein	Green
Carnation by Laboissiere	Floral
Carnation by Mary Chess	Single Flower Floral
Celestiale by Revlon Colourscents	Floral
Chanel No. 5 by Chanel	Aldehydic Modern Blend
Chanel No. 19 by Chanel	Green
Chanel No. 22 by Chanel	Floral
Chantilly by Houbigant	Oriental
Charles of the Ritz by Ritz	Floral
Charlie by Revlon	Fruity Fresh Floral
Chloé by Lagerfeld	Floral
Ciara by Ultima	Oriental
Cie by Shulton	Fruity Fresh Floral
Cinnabar by Estée Lauder	Oriental
Coco by Chanel	Aldehydic Modern Blend
Coriandre by Couturier	Aldehydic Woody Floral
Cristalle by Chanel	Fruity Fresh Floral
Detchema by Revillon Paris	Aldehydic Modern Blend

Dioressence by Dior	Oriental
Diva by Ungaro	Aldehydic Modern Blend
Eau de Gucci by Gucci	Green
Eau de Toilette de Cartier by Cartier	Fruity Fresh Floral
Eau Sauvage by Dior M	Green
Emeraude by Coty	Oriental
Enjoli by Charles of the Ritz	Floral
Epris by Max Factor	Chypre
Estée by Estée Lauder	Floral
Fair Play by Nino Cerruti M	Green
Fantasque by Avon	Floral
Fidji by Guy Laroche	Fruity Fresh Floral
Fire and Ice by Revlon Colourscents	Aldehydic Modern Blend
First by Van Cleef & Arpels	Fruity Fresh Floral
Fleur Bleu by Laboissiere	Fruity Fresh Floral
Fleur de Soire by Laboissiere	Floral
Galanos by Galanos	Oriental
Galore by Germaine Monteil	Floral
Gauloise by Molyneux	Aldehydic Woody Floral
Gian Franco Ferre by Ferre	Floral Blend
Gianni Versace by Versace	Chypre
Giorgio by Giorgio	Floral
Gloria Vanderbilt by Warner	Floral
Halston by Halston	Aldehydic Woody Floral
Hawaiian Jasmine by Langer	Single Flower Floral
Hawaiian Orchid by Langer	Single Flower Floral

Hawaiian White Ginger by Langer	Single Flower Floral
Heliotrope by Mary Chess	Single Flower Floral
Histoire d'Amour by Du Monde	Fruity Fresh Floral
Hope by Frances Denney	Aldehydic Modern Blend
Intimate by Revlon	Chypre
Intrigue by Mary Kay Cosmetics	Oriental
Ivoire de Balmain by Balmain	Green
Jean Louis Scherrer by Scherrer	Fruity Fresh Floral
Jean Naté by Jean Naté	Green
Je Reviens by Worth	Aldehydic Woody Floral
Jontue by Revlon	Floral
Joy by Jean Patou	Floral
Joy Eau de Toilette by Jean Patou	Light and Soft Floral
K de Krizia by Krizia	Fruity Fresh Floral
Keora by Couturier	Oriental
K L by Karl Lagerfeld	Fruity Fresh Floral
L'Air du Temps by Nina Ricci	Floral
Laura Ashley No. 1 by Laura Ashley	Fruity Fresh Floral
Lauren by Ralph Lauren	Fruity Fresh Floral
Le Jardin by Max Factor	Chypre
Love Musk by Jordache	Animalic
Lutèce by Houbigant	Floral
Magical Musk by Toujours Moi	Animalic
Magie Noire by Lancôme	Oriental
Ma Griffe by Myrurgia	Chypre
Maja by Myrurgia	Aldehydic Modern Blend
Maxi by Max Factor	Aldehydic Woody Floral
Mémoire Chérie by Elizabeth Arden	Aldehydic Woody Floral
Metal by Paco Rabanne	Fruity Fresh Floral

Mila Schon by Schon	Chypre
Missoni by Max Factor	Aldehydic Woody Floral
Mist de Mer by Revlon Colourscents	Aldehydic Modern Blend
Molinard de Molinard by Molinard	Fruity Fresh Floral
Muguet by Parfums Prêt-À-Purter	Light and Soft Floral
Musk by Jovan	Animalic
Nikki de Saint Phalle by Cardin	Aldehydic Modern Blend
Nocturnes by Caron	Aldehydic Woody Floral
Nocturnes Cologne Fraîche by Caron	Light and Soft Floral
Norell by Revlon	Fruity Fresh Floral
Nuance by Coty	Aldehydic Woody Floral
Ombre Rose by Jean Charles Brousseau	Aldehydic Modern Blend
Opium by Yves Saint Laurent	Oriental
Oscar de la Renta by Stern	Floral
Paloma Picasso by Picasso	Aldehydic Woody Floral
Parfum de Hermes by Hermes	Floral Blend
Pavi Ellie by Avon	Floral
Pavlova by Payot	Floral
Petale by Revlon Colourscents	Floral
Pheromone by Miglin	Oriental
Private Collection by Estée Lauder	Green
Rafinée by Houbigant	Oriental
Ritz by Charles of the Ritz	Fruity Fresh Floral
Rive Gauche by Yves Saint Laurent	Aldehydic Woody Floral
Royal Secret by Germaine Monteil	Oriental

Ruffles by Oscar de la Renta	Aldehydic Modern Blend
Scoundrel by Revlon	Fruity Fresh Floral
Shalimar by Guerlain	Oriental
Silences by Jacomo	Green
Sinan de Jean Marc Sinan by Jean Marc Sinan	Oriental
Sophia by Coty	Floral
Sportif by Avon	Green
Strategy by Mary Chess	Aldehydic Woody Floral
Tabu by Dana	Oriental
Tapestry by Mary Chess	Aldehydic Modern Blend
Tatiana by Diane Von Furstenberg	Floral
Tawana by Regency	Oriental
Tea Rose by Perfumers Workshop	Floral
Tigress by Fabergé	Aldehydic Woody Floral
Timeless by Avon	Aldehydic Woody Floral
Tuberose by Mary Chess	Sweet Floral
Turbulences by Revillon Paris	Fruity Fresh Floral
Ultima by Revlon	Oriental
Vitabath by Vitabath	Green & Floral
Vivage by Louis Ferauds for Avon	Fruity Fresh Floral
Vivre by Molyneux	Aldehydic Modern Blend
Volcan d'Amour by Diane Von Furstenberg	Aldehydic Woody Floral
Vol de Nuit by Guerlain	Aldehydic Woody Blend
White Lilac by Mary Chess	Light and Soft Floral
White Linen by Estée Lauder	Aldehydic Modern Blend
White Shoulders by Evyan	Floral
Wind Song by Prince Matchabelli	Floral
With Pleasure by Caron	Light and Fresh Floral

Y by Yves Saint Laurent	Green
Youth Dew by Estée Lauder	Oriental
Ysatis by Givenchy	Aldehydic Woody Blend

One of my favorite stories is about a client who, because perfumes didn't seem to last on her, chose to wear oriental scents. She detested oriental fragrances but wore them because she thought they lasted longer than anything else. She discovered her basic fragrance category was fruity fresh floral. I showed her how to choose fragrances from that category more suitable for throw-offs. She vowed to never again buy oriental scents or wear them even when they were gifts.

If you throw off scents, it *does* matter which kind of fragrance you select. Some perfumes are heavier than others. Heavier refers to the kinds of ingredients found in a fragrance. We know orientals, like Youth Dew of Estée Lauder, are heavier than greens like Chanel No. 19, but some greens, like Lauder's Aliage, are heavier than other greens, like Jean Naté's. Throw-offs should wear heavier scents from the category of their choice because they will last longest on their skin.

The Name-Brand Fragrance Guide lists more than 300 perfumes by basic fragrance category. In each category, those fragrances that have heavier properties are clearly identified with an H. Throw-offs should choose from among those fragrances so marked.

You might be surprised to find a few men's fragrances included in the guide. Men's scents are made with lavender, carnation, orange, oakmoss, clove, sandalwood —the *same* ingredients found in women's fragrances. Eau Sauvage of Christian Dior was one of the first fragrances marketed for men but meant to appeal to women as well. Men's colognes are less expensive per ounce than women's fragrances. Because your body chemistry will affect the fragrance, it won't smell the same way on you as it does on a man.

The Name-Brand Fragrance Guide is a tool to be

used in selecting your fragrances. Classified by basic fragrance category, the name brands are listed alphabetically. The guide will become indispensable; it saves time and money and limits confusion.

THE NAME-BRAND FRAGRANCE GUIDE

Fragrances are listed alphabetically by Basic Fragrance Category. Product name is listed first, then manufacturer. Heavier scents are marked with an H. (Men's fragrances are marked with an M.)

The Green Basic Fragrance Category

H   Aliage by Estée Lauder
H   Armani Eau Pour Homme by Giorgio Armani   M
    Body Basics by Elizabeth Arden
    Calvin Klein by Calvin Klein
    Capucci Pour Homme by Capucci   M
    Chanel No. 19 by Chanel
H   Devin by Aramis   M
    Drakkar by Guy Laroche   M
    Eau de Gucci by Gucci
    Eau de Guerlain by Guerlain
    Eau de Hermès by Hermès
    Eau de Imperiale by Guerlain   M
    Eau de Patou by Patou
    Eau de Toilette de Cartier by Cartier
    Eau Fresh by Jovan
    Eau Sauvage by Christian Dior   M
    Equipage by Hermès
    Fair Play by Nino Cerruti   M
    Givenchy III by Givenchy
    Graffiti by Capucci
H   Grey Flannel by Geoffrey Beene   M
    Halston 1–12 by Halston   M
    Herbessence by Helena Rubenstein
    Inoui by Shiseido
    Iris de Fête by Jontue-Revlon
H   Ivoire de Balmain by Balmain
    Jean Naté by Jean Naté

H Le Must de Cartier by Cartier
  Love's Fresh Lemon by Chattem
  Monsieur de Givenchy by Givenchy M
  Murasaki by Shiseido
  Natural de Myrurgia by Myrurgia
  Oscar de la Renta Pour Lui by
   Oscar de la Renta M
  Pierre Cardin For Men by Pierre Cardin M
H Private Collection by Estée Lauder
  Sarogasso by Perfumers Workshop
H Silences by Jacomo
  Soft Rain by Love
  Sportif by Avon
  Spring Green by Vitabath
  Tangy Lemon by Vitabath
  Trophée by Lancôme M
H Vent Vert by Balmain
  Vertable by Guerlain M
  Y by Yves Saint Laurent

Other Alternatives

Green scents are often mixed with basic scents from other categories. I have included a listing of fragrances that contain green ingredients for alternating your fragrance choices.

  Amazone by Hermès—green fruity fresh floral
  Cabochard by Grès—green chypre
  Casaque by Jean d'Albret—green fruity fresh
   floral
H Gianni Versace by Gianni Versace—green chypre
H Jean Louis Scherrer by Jean Louis Scherrer
   —green fruity fresh floral
H Ma Griffe by Myrurgia—green chypre

The Fruity Fresh Floral Basic Fragrance Category

  Amazone by Hermès
  Anais Anais by Cacharel
  Anne Klein Parfum by Anne Klein

H   Armani by Giorgio Armani   M
    Blason by Myrurgia
    Charlie by Revlon
    Cie by Shulton
    Cristalle by Chanel
    Diorella by Dior
    Diorissimo by Dior
    Fidji by Guy Laroche
    First by Van Cleef & Arpels
    Fleur Bleu by Laboissiere
    Fleurs d'Orlane by Jean d'Albret
    Histoire d'Amour by Du Monde
H   Jean Louis Scherrer by Jean Louis Scherrer
    K de Krizia by Krizia
    K L by Karl Lagerfeld
    Laura Ashley No. 1 by Laura Ashley
    Lauren by Ralph Lauren
    Le Temps d'Aimer by Alain Delon
H   Madame de Carven by Carven
    Metal by Paco Rabanne
    Molinard de Molinard by Molinard
    Muguet des Bois by Coty
H   Norell by Revlon
    Quartz by Molyneux
    Ritz by Charles of the Ritz
    Scoundrel by Revlon
    Sex Appeal by Jovan
    Touch of Class by Fabergé
H   Turbulences by Revillon Paris
    Vivage by Louis Ferauds for Avon
    Zarolia by Maitland Phillipe
    Zen by Shiseido

The Floral Basic Fragrance Category

There are more floral fragrances manufactured than
any other kind. I have subdivided this category into
four parts: sweet florals, light and soft florals, floral
blends, and single flower florals. To find your category,

select a favorite floral fragrance and observe its sub-category.

Sweet Florals

    Adolpho by Frances Denney
H   Babe by Fabergé
    Bill Blass by Revlon
    Blue Grass by Elizabeth Arden
    Calais by Mary Kay Cosmetics
    Celisse by Dana
    Cerissa by Ultima
    Charles of the Ritz by Revlon
H   Chloé by Karl Lagerfeld
    Ebullience by Neiman-Marcus
H   Echo by Borghese
    Enjoli by Charles of the Ritz
    Estée by Estée Lauder
    Fleur de Soie by Laboissiere
    Fleurs de Rocaille by Caron
H   Fracas by Robert Piguet
    Gloria Vanderbilt by Warner
    Great Lady by Evyan
    Hawaiian Gardenia by Langer
    Hawaiian Plumeria by Langer
    Island Gardenia by Jovan
    Jontue by Revlon
    L'Heure Bleu by Guerlain
    Madeline de Madeline by Madeline Mono
    Michelle by Balenciaga
    Moon Drops by Revlon
    Most Precious by Evyan
    Oscar de la Renta by Stern
    Pavlova by Payot
    Tatiana by Diane Von Furstenberg
    Tuberose by Mary Chess

Light and Soft Florals

    Always Alluring/Impulse by Lever Bros.

Après L'Ondée by Guerlain
Beret by Louis Phillipe
Cardin by Pierre Cardin
Celestiale by Revlon Colourscents
Chanel No. 22 by Chanel
Delightfully Daring/Impulse by Lever Bros.
Embracing by English Leather
Emma by Laura Ashley
En Fleur by Alissa Ashley
Fleurs de Fleurs by Nina Ricci
Flora Danica by Swank
Instantly Innocent/Impulse by Lever Bros.
Jardins de Bagatelle by Guerlain
Joy Eau de Toilette by Jean Patou
L'Air du Temps by Nina Ricci
Laura Biagiotti by Laura Biagiotti
Le Dix by Balenciaga
Les Fleurs by Alissa Ashley
Lotus de Nuit by Jontue-Revlon
Muguet by Parfums Prêt-À-Porter
Nocturnes Cologne Fraîche by Caron
Possibly Playful/Impulse by Lever Bros.
H   Pour un Homme de Caron by Caron   M
H   Premiere by Jean Charles Castelbajac
Princess Livia by Princess Livia
Rare Blossoms/Body Flowers by Gillette
Rose de Mai by Jontue-Revlon
Spring Petals/Body Flowers by Gillette
Suddenly Sassy/Impulse by Lever Bros.
Sweet Honesty by Avon
Wedding Bouquet by Floris
White Lilac by Mary Chess
With Pleasure by Caron

Floral Blends

Andiamo by Borghese
Avenir by Mary Kay Cosmetics
Blazer by Helena Rubenstein

Courant by Helena Rubenstein
Di Borghese by Borghese
H   Fantasque by Avon
H   Filly by Capucci
Forever Krystle by Charles of the Ritz
Galore by Germaine Monteil
Gian Franco Ferre by Ferre
Giorgio by Giorgio
H   Gucci by Gucci
Jordache Woman by Jordache
H   Joy by Jean Patou
Lady by Jovan
L'Aimant by Coty
Laughter by Yardley
La de Givenchy by Givenchy
Lutèce by Houbigant
H   Lys Bleu by Prince Henri Pierre d'Orleans
Malmaison by Floris
Marilyn Monroe by CFT
Millefleurs by Crabtree and Evelyn
Natchez by Margaret Hodge
Normandie by Jean Patou
Paradis by Saks Fifth Avenue
Parfum de Hermes by Hermes
Paris by Yves Saint Laurent
Pascalle by Kensington
H   Patou 1000 by Jean Patou
Pavi Elle by Avon
Petale by Revlon Colourscents
Prelude by Balenciaga
Quelques Fleurs by Houbigant
Sophia by Coty
Symbiose by Stendahl
H   Tapestry by Mary Chess
VSP by Jovan
H   White Shoulders by Evyan
Wind Song by Prince Matchabelli
Yendi by Capucci

Single Flower Florals

    Aqua de Lavande by Myrurgia
    Carnation by Laboissiere
    Carnation by Mary Chess
    Gardenia by Laboissiere
    Gardenia/French Garden Flowers by Houbigant
    Hawaiian Jasmine by Langer
    Hawaiian Orchid by Langer
    Hawaiian White Ginger by Langer
    Heliotrope by Mary Chess
    Iris by Laboissiere
    Jasmine by Laboissiere
    Jasmine/French Garden Flowers by Houbigant
    Lilac by Laboissiere
    Lilac/French Garden Flowers by Houbigant
    Misty Tea Rose by Jovan
    Muguet by Laboissiere
    Night Blooming Jasmine by Jovan
    Rose Imperial by Laboissiere
H   Tea Rose Extrait by Perfumers Workshop
H   Tea Rose/French Garden Flowers by Houbigant
    Tuberose by Laboissiere
    Violette by Laboissiere

The Aldehydic Modern Blend
Basic Fragrance Category

    Albert Nipon by Albert Nipon
    Aphrodisia by Fabergé
H   Arpège by Lanvin
    Azurée by Estée Lauder
    Carnation by Laboissiere
    Champagne by Germaine Monteil
    Chanel No. 5 by Chanel
    Coco by Chanel
    Detchema by Revillon Paris
    Diva by Ungaro
    Fiamo by Revlon
    Fire and Ice by Revlon Colourscents

H   Habit Rouge by Guerlain
H   Hope by Frances Denney
     Infini by Caron
     L'Interdit by Givenchy
H   Madame Rochas by Madame Rochas
H   Maja by Myrurgia
     Mist de Mer by Revlon Colourscents
H   Mystère by Madame Rochas
     Nikki de Saint Phalle by Pierre Cardin
     Odyssey by Avon
H   Ombre Rose by Jean Charles Brousseau
H   Pois de Senteur by Caron
     Régine by Continental
     Ruffles by Oscar de la Renta
     Sand and Sable by Regency
     Tapestry by Mary Chess
     Tuxedo by Ralph Lauren
H   Vivre by Molyneux
     White Linen by Estée Lauder

The Aldehydic Woody Floral
Basic Fragrance Category

     Alexandra by Alexandra de Markoff
     Ambre D'Or by Parfums Prêt-À-Porter
     Amouage by Asprey
     Angel Fire by Mary Kay Cosmetics
     Aviance by Prince Matchabelli
     Bijan by Bijan   M
H   Bois des Iles by Chanel
     Calandre by Paco Rabanne   M
     Calèche by Hermès
     Chamade by Guerlain
     Coriandre by Couturier
     Essence Rare by Houbigant
     Fleurs des Bois by Laboissiere
H   Gauloise by Molyneux
     Halston by Halston
H   Halston Night by Halston

Heaven Scent by Mem
H Je Reviens by Worth
Love's Baby Soft by Chattem
H Maderas by Myrurgia
Maxi by Max Factor
Mémoire Chérie by Elizabeth Arden
Missoni by Max Factor
H My Sin by Lanvin
H Nahema by Guerlain
H Nocturnes by Caron
Nuance by Coty
Oleg Cassini by Jovan
Paloma Picasso by Paloma Picasso
Rive Gauche by Yves Saint Laurent
Snow Buds/Body Flowers by Gillette
Strategy by Mary Chess
H Tigress by Fabergé
Timeless by Avon
Touché by Jovan
H Volage by Neiman-Marcus
Volcan d'Amour by Diane Von Furstenberg
Vol de Nuit by Guerlain
Ysatis by Givenchy

The Chypre Basic Fragrance Category

Andron by Jovan
Azarro by Louis Azarro
H Cabochard by Grès
Cachet by Prince Matchabelli
H Cachet Noir by Prince Matchabelli
Chimère by Prince Matchabelli
H Ciao by Houbigant
Daydreams by Maybelline
H delys by Charles Lamaine
H Epris by Max Factor
H Femme by Madame Rochas
Geminesse by Max Factor
Gianni Versace by Gianni Versace

H   Intimate by Revlon
H   Lanvin for Men by Lanvin   M
     Laura Ashley No. 2 by Laura Ashley
     Le Jardin by Max Factor
     Ma Griffe by Myrurgia
H   Mila Schon by Mila Schon
H   Miss Dior by Christian Dior
     Mitsouko by Guerlain
     Sculptura by Jovan
     Senchal by Charles of the Ritz
H   7e Sens by Sonia Rykiel

The Oriental Basic Fragrance Category

     Ambermist by Avon
H   Ambush by Dana
     Amun by Canteline
H   Aromatics Elixir by Clinique
H   Bal à Versailles by Jean Desprez
H   Bellodgia by Caron
H   Black Sapphire by Revlon Colourscents
     Cafe by Parfums Cafe
H   Chantilly by Houbigant
H   Ciara by Ultima
     Cinnabar by Estée Lauder
     Confidence by Laboissiere
     Dioressence by Christian Dior
H   Emeraude by Coty
H   Enigma by Alexandra de Markoff
H   Enjoli Midnight by Charles of the Ritz
     Galanos by Galanos
H   Interlude by Frances Denney
     Intrigue by Mary Kay Cosmetics
     J'ai Ose by Guy Laroche
     Jicky by Guerlain
     Keora by Couturier
     L'Origan by Coty
H   Magie Noire by Lancôme
     New York! New York! by Parfums Vionnet

Ninja by Parfums Coeur
H   Nuit de Noël by Caron
H   Opium by Yves Saint Laurent
Or de Torrente by Torrente
Pheromone by Miglin
Rafinée by Houbigant
H   Royal Secret by Germaine Monteil
H   Sinan de Jean Marc Sinan by Jean Marc Sinan
Shalimar by Guerlain
H   Sheherazade by Jean Desprez
H   Tabac Blond by Caron
H   Tabu by Dana
Tawana by Regency
Ultima by Revlon
Vertiver by Carven   M
Vertiver by Guerlain   M
Woodhue by Fabergé
H   Youth Dew by Estée Lauder

The Animalic Basic Fragrance Category

Autumn Musk/Body Flowers by Gillette
H   Aviance Night Musk by Prince Matchabelli
Civet by Alissa Ashley
Love Musk by Jordache
Magical Musk by Toujours Moi
H   Midnight Musk by Bonne Bell
Musk by Alissa Ashley
Musk by Jovan
Musky Jasmine by Chattem
H   Mysterious Musk/Impulse by Lever Bros.
H   Skin Midnight Musk by Bonne Bell
Skin Musk by Bonne Bell
Soft Musk by Avon
H   Tigress Musk by Fabergé
Whisper of Musk by Jovan
H   Wild Musk by Coty
Woman by Jovan

PERFUME APPLICATION GUIDELINES

Now that you have chosen a fragrance appropriate to *you*, it is time to learn how to wear it properly.

The late Coco Chanel answered the question, "Where can I wear perfume?" by replying, "Wherever you want to be kissed." Another way of answering the question is that perfume should be applied at your pulse points.

The question of how much to apply is more complicated. First of all, if you are using a perfume spray, always use it from a distance of at least eight inches from your body. Spraying perfume from too close a range produces the same effect as spraying paint from too short a distance: it will run and be distributed unevenly.

Each time you apply perfume, use either three measured sprays or several drops (if you are applying perfume from a bottle). The amount of perfume you apply each time doesn't vary from heavy to light; what does change is how *many* times a day you apply the perfume. Remember that although the applications may not seem heavy enough for you, by the time you have been wearing a perfume for a while your sensitivity to it tends to dull. Other people will be able to smell it even though you can't. Follow the application guide and try not to succumb to the temptation of spraying yourself with perfume all the time. (This guide applies to all fragrance strengths.)

Perfume Application Guide

This is an application guide for people who don't throw off scent.

* Apply perfume four or five times a day.
* First apply it before work.
* Use it for a quick pick-me-up before lunch.
* Use it as a refresher after work.
* Spray yourself lightly with it before going to bed.

Remember, by the way, that it is a great mistake to spray yourself with a perfume a few seconds before a glamorous date or a vital meeting. The fragrance is very strong when you first put it on and may alienate the person you are about to meet. Instead, try to apply it at least five minutes before the meeting.

Throw-Off Application Guide

This is for everyone who throws off scent and scored more than 20 in the scientific test.

* Use the Perfume Application Guide above as a base guide in applying your fragrance.
* If you are using fragrances marked H, don't make the mistake of applying too much at one time. Remember, even if you can't smell the scent, others around you can.
* Layer other fragrance products for a longer lasting scent. Choose cremes, lotions, powders, or bath oils.
* Carry cologne sticks, scented towelettes, and touch-tip fragrance wands in your purse or desk drawer for a light, inexpensive way to renew your fragrance.

Remember that many of the perfumes graded H are heavy fragrances like Youth Dew, Aromatics Elixir, and Intimate, which all have amber bases. They are chock full of resins and gum, which interact with the fixatives of a fragrance. Fragrance with a very strong amber base note lasts longer and therefore doesn't throw off so easily and doesn't need to be applied frequently.

Remember, too, that climate affects perfume. You will find that you need to apply more in cold weather, when it evaporates quickly.

CHAPTER · 7

Step Four: Building A Perfume Wardrobe

Today's woman is multidimensional. She is no longer merely the frame to some man's picture. She has an identity of her own. And that identity has many aspects. Today's woman can be a mother and a housewife, a businesswoman and a femme fatale, and athlete and a sophisticated hostess. I strongly believe that as a woman's role changes, so should her fragrance.

Once you have selected your favorite perfume category, Step Four is designing your own fragrance wardrobe. I believe every woman needs one. I am not going to advise you all to rush out and buy ten $150 bottles of perfume. I want to give you enough alternatives to prevent your wasting money and making wrong decisions about your own perfume style. Perfume should be chosen and worn as carefully as clothes are. Just as you shouldn't wear the same outfit to a cocktail party as you would to a baseball game, you shouldn't wear the same fragrance either.

Building a perfume wardrobe is easier than it sounds. The lifestyle you lead becomes the foundation for the type of fragrances you will need. I believe women need to cover four main areas of lifestyle in making their fragrance selections: everyday wear, sport wear, evening wear, and intimate wear.

Take a look inside your closet. Clothes can expose the kind of lifestyle you lead. How many suits and dresses can you find? How about dress pants, blouses, and sweaters? Are there any jogging pants, T-shirts, or tennis shoes? How often do you wear sport clothes versus casual or dressy clothing? Do you see a lot of evening clothes hanging on the racks? The kind of clothes you possess reveals the kind of lifestyle you prefer.

LIFESTYLES AND FRAGRANCES

* <u>Everyday wear</u> encompasses home and work. We, as women, have learned how to dress in the workplace. Business suits, low-heeled pumps, and well-groomed hair have become the new standards we measure ourselves by. Why ignore your fragrance? The wrong choice can do as much harm as wearing a high-on-the-thigh slit skirt and halter top to the office. For example, after I spoke at one businesswoman's luncheon, an executive of a computer firm asked for a consultation. She recently had been passed over for a promotion and she was re-evaluating her personal style. She wanted to include a change from the oriental fragrance she usually wore to work. I advised her to choose a crisp, fresh scent that, worn in the work force, could project an image of efficiency, thoroughness, and competence. Fragrance is a type of body language that signals to the world who we are and how we want to be perceived. Clearly, for this executive, the wrong message was being communicated.

* <u>Sport wear</u> has prompted manufacturers to make their already popular perfumes in lighter concentrations. More and more women are participating in some form of physical exercise, and they want to wear a fragrance while doing it. Perfume and perspiration do not mix. Green and fruity fresh florals are appropriate for tennis, jogging, or working out with Jane Fonda. If, however, extremely heavy or sweet floral scents, rich orientals or spicy aldehydic modern blends are your unshakable favorites, choose a lighter concentrate of your favorite scent. Another way to achieve a fragrant smell while practicing your aerobic routine is the use of lotions, body splashes, cologne sticks, or scented towelettes—you don't have to spray on a fragrance in order to smell nice!

* <u>Evening wear</u> to me means parties, dates, or perhaps an evening or dinner on the town. It is a romantic, seductive, and personal time for choosing fragrance. Now is when it's right to wear the sweet floral, rich oriental, or spicy aldehydic modern blend. When layered with other fragrance products (lotions, milk baths, powders), eau de toilettes are excellent choices for evening. If your evening out is rather special, or black tie, then by all means wear your most elegant perfume. Fragrances from the perfume family remind me of silk and pearls and being, as they say, all dressed up!

* <u>Intimate wear</u> is the time for all-out glamour and seduction—but not necessarily in a fragrance choice. It might be surprising to you that I don't come on strong for a fragrance from the perfume family for intimate wear. I think perfume is more suitable for evening wear, when you want and need a fragrance to last through the rounds of cocktails, dinner, and dancing. Drenching yourself in the strongest concentrated form of fragrance available for an intimate evening can be an immediate turn-off. Recently, *Glamour* magazine surveyed men about their fra-

grance preferences. A whopping 85 percent of them reported that they liked a hint of perfume and 10 percent said the greatest turn-off was when a woman wore too much fragrance. Any products from the eau de toilette family and a few products from the perfume family (in lighter categories, such as green and fruity fresh floral) work well for intimate wear. Bath oils, applied on pulse points, can be substituted for perfume. In addition to the dollar savings, the oil will moisturize and soften your skin. Remember, though, don't go overboard in applying the stronger concentration of fragrance.

BUILDING YOUR FRAGRANCE WARDROBE

Many women then ask me the question, "How many different fragrances will I need in my fragrance wardrobe?" To begin, I recommend *two* fragrances, in a total of *three* concentrations. Eventually, a well-rounded fragrance wardrobe will include five to six different fragrances and as many fragrance products as you enjoy using.

For everyday wear, I believe you should own two fragrances (for variety), in either eau de cologne or eau de toilette. Sport wear requires one light fragrance in an eau de cologne or body splash, body refresher, etc. For evening wear, a perfume or eau de toilette are perfect. I like to be a little extravagant in the intimate-wear area because I don't want to smell the same way each time I'm in a romantic mood. So, I would suggest two different fragrances; eau de toilettes, bath oils, or light perfumes make excellent choices. Eau de parfum toilette from the perfume family is another good selection.

Now, let's begin your fragrance wardrobe by making just three purchases. Use the following chart to take another look at my wardrobe recommendations.

Lifestyle Area	Number of Fragrances Recommended	How to Achieve
Everyday wear	2	EDT* in Fragrance X EDC* in Fragrance Y
Sport wear	1	EDC in Fragrance Y
Evening wear	1	EDT or Perfume in Fragrance X
Intimate wear	2	EDT or EDPT* in Fragrance X Perfume or Bath Oil in Fragrance Y

* EDT = Eau de Toilette
EDC = Eau de Cologne
EDPT = Eau de Parfum Toilette

The eau de toilette in fragrance X performs triple duty for everyday, evening, and intimate wear. The eau de cologne in fragrance Y does double time for everyday and sport wear (light enough for both the office and the exercise gym). The bath oil in fragrance Y qualifies for both evening and intimate wear. Two fragrances, three concentrations—see how simple?

After acquiring three concentrations, you should next invest in other fragrance products such as body cremes, dusting powders, or cologne sticks. These products help preserve a faint lingering of scent that lasts all day long. They can also stretch a fragrance wardrobe by allowing for less usage of perfumes, toilettes, and colognes. As your budget allows, begin to add one new fragrance and a few of its products at a time. Slowly build your wardrobe until you have a well-rounded selection of fragrances from which to choose.

I personally try to match my fragrance choices to what I am doing. I strongly believe that the messages we convey through our fragrance must tally with our personality and the situation we are in. I have two hard and fast rules concerning fragrance-wear:

1. No heavy, seductive fragrances should be worn for business.
2. No heavy or sweet fragrances should be worn for sports activities.

Those of you who throw off scent might be confused here. I've told you to choose among heavy or H-marked fragrances because of your body chemistry. Remember how I also told you about layering products? These, then, become the occasions to layer several of the less concentrated products in your fragrance. Yes, you *can* wear fragrances from your favorite category, but at your next business meeting, you don't want to overcome those around you with heavy perfumes.

As I've previously stated, perspiration and perfumes don't mix. When you perspire, your body secretes certain chemicals and those chemicals can interact with scent, drastically changing its character. Fragrances from the green category are perfect for sports. They are fresh, clean, crisp, and have an uplifting effect. But I do realize some women don't like green fragrances. For them, wearing a scent from their favorite category is permissible, as long as it is in the lightest concentration available. Lotions, frictions, body splashes, or cologne sticks make excellent sports fragrance choices.

A BASIC WEARING GUIDE

Some women meet friends for lunch, visit a relative in the nursing home, or fly to the Rockies on their vacation. Other women go to work every day, serve as volunteers, go to garage sales on weekends, or spend their vacation each summer at the beach. When a woman calls for an individual fragrance consultation, we explore every aspect of her lifestyle, including day-to-day living.

Everyday Living

The object here is to teach you which fragrances work

best in day-to-day, real-life situations. Most of our daily routines include some of the following:

Meeting friends for lunch
Fulfilling doctor or dentist appointments
Visiting friends or relatives in nursing homes or hospitals
Shopping
Working and attending business meetings
Performing volunteer work
Running errands—grocery buying, picking up the kids, etc.
Cleaning house
Preparing meals

I'm sure you can add more to the list by reviewing your own daily routine. Fragrance concentrations to choose for day-to-day living are mostly eau de colognes and a few selective eau de toilettes. Can you imagine yourself pushing a cart in the grocery store wearing a glamorous, enticing, and heavy perfume? Lighter, unobtrusive scents work best around older relatives, co-workers, and children. Sweet smells tend to nauseate sick people.

Sports Activities

Are you a spectator or a participant in sports activities? I love attending football games during the cool fall months and swimming every day during the hot summertime. The colder seasons make fragrance evaporate faster whether you are sitting in bleachers or racing downhill on skis. If you like the green category fragrances for sports, choose from those marked with an H. This advice is for *all* body chemistry types. Green category fragrances are basically lighter scents than other categories. Men's colognes make excellent winter sports fragrances as they seem to last longer on the skin than women's fragrances. Eau de colognes and eau de toilettes serve winter sports admirably. Also, stronger concentrations from a lighter fragrance category (green, fruity fresh floral) enable a scent to have staying power in the colder months.

When you are at the beach or lying around a swimming pool, don't wear fragrance at all. Scents worn in direct sunlight can stain your skin. Furocoumarins, found in fragrances, react with the sun, causing exaggerated sunburn, which means that your skin has become photosensitive or has a phototoxic reaction. You might think you have a normal sunburn, but within one or two days you will develop blistering, hives, some swelling so severe that you may require medical treatment. You will then become extremely sensitive to sunlight for weeks or months.

Travel Time

Most people love to travel, and I'm no exception! But I do not like to take a seat on a crowded train or plane and find the person next to me reeking of scent, especially something heavy such as an oriental or animalic fragrance, even though I love oriental scents! Traveling has a tendency to fray nerves and exhaust and stiffen bodies. Fragrance can give you a psychological lift. Fresh-smelling scents will create the feeling of cleanliness and tidiness amid the chaos and rush of traveling. Choose only light, crisp fragrances when you travel. Eau de colognes, body splashes, body fresheners, eau fraiche, and towelettes not only smell good, they refresh! High altitudes decrease a fragrance's scent-ability. If it's off to the Rockies for you, select stronger concentrations in lighter categories. For example, an eau de parfum toilette from the green, fruity fresh floral, or light and soft floral categories will keep you smelling good when you are 8,000 feet above sea level.

Elegant Occasions

I attend four or five black-tie functions every year and I adore dressing for these events. My silk and satin lingerie comes out of the bottom drawer of my lingerie chest, and my diamonds (I'll admit they're small, but they *are* good!) are cleaned and clipped to my ears. I take my most elegant bottle of pure perfume and apply it slowly on each pulse point.

Other special occasions can include the ballet or symphony, weddings, dinner parties, evenings on the town, christenings—the list could go on and on. You need a fragrance that will last throughout the entire occasion and smell as strong as when you first applied it. Fragrances from the perfume family are perfect for the demands of the elegant occasion. Perfumes do not have to be expensive to be real. They are available in all price ranges.

Romance and Seduction

Romance can imply a stroll in the garden, an evening spent curled up in front of the fire, or a candlelight dinner for two. Fragrances can sometimes overpower a romantic mood. Most men are turned off by loud, suggestive scents.

A fragrance should convey, gently, your romantic mood, and it should also be an extension of your femininity. Bold, strong fragrances are too heavy for romance and seduction. Better choices would be an eau de toilette along with layerings of other fragrance products. Again, an eau de parfum toilette, which is more concentrated, is also appropriate if it is from one of the lighter fragrance categories.

THE WEARING CHART

Use the following chart to tell at a glance which fragrance concentrations are the best choices for your lifestyle areas.

Lifestyle Area	Eau de Cologne	Eau de Toilette	Eau de Parfum
Everyday	✓	✓	
Office and Work	✓	✓	
Sports	✓		✓ –Winter months
Traveling	✓		✓ –High altitudes
Elegant Occasions			✓
Romance and Seduction		✓	✓ –Lighter categories

Certain fragrances are more appropriate for certain ages as well. My grandmother, for example, loves the light, flowery blends best of all. She feels that heavier scents are wrong for elderly women. Younger girls should refrain from wearing fragrances that are too heavy or too grown-up. A young lady (ten and up) should have her own fragrance. Look for powdery or light, citrus-type scents. Let her choose from among those you have previously screened. Consulting with the child helps her develop her own fragrance sense.

Take clues from your clothes. Plastic thong sandals don't work with linen business suits, and three-inch-high heels look out of place when worn with walking shorts. Don't be the woman who wears a body splash with her evening gown or the one who wears parfum concentrate with her jogging shorts. And don't be the woman who wears the latest designer fragrance *everywhere*. The current trend might not be right for you. Polish your total look with fragrances chosen to correspond with the image you are projecting, your lifestyle, and your personality!

PERFUME SUGGESTIONS FOR ALL OCCASIONS

The six charts that follow provide my specific suggestions for fragrances appropriate to the various aspects of a woman's busy life. You may discover your own personal fragrance choices in the lists or you may find inspiration for broadening your perfume wardrobe.

Note: The following abbreviations are used: FFF = Fruity Fresh Floral; G = Green; F = Floral; AMB = Aldehydic Modern Blends; AWF = Aldehydic Woody Florals. See the explanation of fragrance categories on pp. 32–35.

PERFUME SUGGESTIONS FOR EVERYDAY ACTIVITIES

Category	Name	Appropriate Products
G	Calvin Klein by Klein	Eau de Cologne Family
G	Chanel No. 19 by Chanel	Eau de Toilette Family
FFF	Anais Anais by Cacharel	Powders
FFF	Cie by Shulton	Lotions
Light and Soft F	Rare Blossoms/Body Flowers by Gillette	Bath/Shower Products
Light and Soft F	Emma by Laura Ashley	Body Splash
Sweet F	Enjoli by Charles of the Ritz	
Sweet F	Oscar de la Renta by Stern	
F Blends	Galore by Germaine Monteil	
F Blends	Wind Song by Prince Matchabelli	
Single F	Carnation by Laboissiere	
Single F	Tea Rose/French Garden Flowers by Houbigant	
AMB	H Hope by Frances Denney	Everyday activities encompass running errands, grocery shopping, picking up the kids, etc. The fragrances you select should not be strong but should reflect something fresh, romantic, a bit spirited. Any fragrances from the green category or the light and soft floral category will work well here.
AMB	Ruffles by Oscar de la Renta	
AWF	Rive Gauche by Yves Saint Laurent	
AWF	Maxi by Max Factor	
Chypre	Laura Ashley No. 2 by Laura Ashley	
Chypre	Ma Griffe by Myrurgia	
Oriental	Dioressence by Dior	
Oriental	Aromatics Elixir by Clinique	

PERFUME SUGGESTIONS FOR OFFICE, WORK, AND BUSINESS MEETINGS

Category	Name	Appropriate Products
G	Eau Sauvage by Christian Dior	Eau de Cologne Family
G	H Aliage by Estée Lauder	Eau de Toilette Family
FFF	Laura Ashley No. 1 by Laura Ashley	Powders
FFF	Charlie by Revlon	Body Splash
Sweet F	Tatiana by Diane Von Furstenberg	Lotions
Sweet F	Bill Blass by Revlon	Cremes
F Blends	Blazer by Helena Rubenstein	Bath/Shower Products
F Blends	Millefleur by Crabtree & Evelyn	Scented Deodorant
Single F	Iris by Laboissiere	Towelettes (Keep them at the office!)
Single F	Lilac/French Garden Flowers by Houbigant	
Light and Soft F	Beret by Louis Phillipe	
Light and Soft F	L'Air du Temps by Nina Ricci	
AMB	White Linen by Estée Lauder	
AMB	Albert Nipon by Albert Nipon	
AWF	Aviance by Prince Matchabelli	
AWF	Oleg Cassini by Jovan	
Chypre	Cachet by Prince Matchabelli	
Chypre	Daydreams by Maybelline	
Oriental	Keora by Couturier	
Oriental	Cinnabar by Estée Lauder	

I deliberately left out animalic scents because they are not appropriate to the working environment. I don't recommend oriental scents either; but for women who insist on wearing heavier fragrances, I have included two of the lightest orientals. Women who throw off scent and need H-marked fragrances can layer light fragrance products to make the scent last.

PERFUME SUGGESTIONS FOR SPORTS

Category	Name	Appropriate Products
G	Body Basics by Elizabeth Arden	Eau de Cologne Category
G	Eau Fresh by Jovan	Powders
G	Jean Naté by Jean Naté	Lotions
G	Love's Fresh Lemon by Chattem	Towelettes
G	Natural by Myrurgia	Scented Deodorants
G	Spring Green by Vitabath	Body Splash
G	Tangy Lemon by Vitabath	Bath or Body Scrubs
FFF	Cristalle by Chanel	Gels
FFF	Diorella by Dior	Touch-on Applicators
Sweet F	Blue Grass by Elizabeth Arden	
Light and Soft F	Instantly Innocent/Impulse by Lever Bros.	
Light and Soft F	Spring Petals/Body Flowers by Gillette	
Light and Soft F	Sweet Honesty by Avon	
F Blends	Laughter by Yardsley	
F Blends	La de Givenchy by Givenchy	
Single F	Aqua de Lavande by Myrurgia	
Single F	Lilac by Laboissiere	

Remember, perfume and perspiration do not mix. Yes, it is possible to wear fragrances from the other categories, but I recommend wearing their additional products if you are participating in a sport. For example, cool towelettes, refreshing body lotions, soothing powders, moisturizing bath sprays, each in your particular scent, perhaps layered, applied before your exercise routine or tennis game. You will be able to smell good and work hard at the same time!

PERFUME SUGGESTIONS FOR TRAVELING

Category		Name	Appropriate Products
G	H	Vent Vert by Balmain	Eau de Cologne Family
G	H	Aliage by Estée Lauder	Eau de Perfume (ONLY in high altitudes, lightly applied)
G		Body Basics by Elizabeth Arden	Towelettes
		Eau Fresh by Jovan	Scented Deodorants
G		Jean Naté by Jean Naté	Body Splashes
G		Love's Fresh Lemon/Love by Chattem	Solids
G		Soft Rain by Love	Powders
FFF		Diorissimo by Dior	Silks
FFF		Muguet des Bois by Coty	Lotions
Light and Soft F		Les Fleurs by Alissa Ashley	Bath/Shower Products
Light and Soft F		Flora Danica by Swank	
F Blends		Laughter by Yardley	
Sweet F		Blue Grass by Elizabeth Arden	
Sweet F		Tatiana by Diane Von Furstenberg	
Single F		Aqua de Lavande by Myrurgia	
Single F		Carnation by Laboissiere	If your category is not listed, then select other types of products in your favorite category, such as powders, lotions, towelettes. Traveling time is made easier with light, refreshing fragrances.
AMB		Azurée by Estée Lauder	
AMB		White Linen by Estée Lauder	
AWF		Love's Baby Soft by Chattem	
AWF		Heaven Scent by Mem	

PERFUME SUGGESTIONS FOR ELEGANT OCCASIONS

Category		Name	Appropriate Products
G	H	Ivoire de Balmain by Balmain	Eau de Perfume Family
G	H	Silences by Jacomo	Lotions
FFF	H	Jean Louis Scherrer by Scherrer	Cremes
FFF	H	Norell by Revlon	Powders
Sweet F	H	Fracas by Robert Piguet	Bath Oils
Sweet F		Estée by Estée Lauder	Solids (to carry in a solid compact)
Light and Soft F		Princess Livia by Princess Livia	Purse Spray
Light and Soft F		Premiere by Jean Charles Castelbajac	
F Blends	H	Joy by Jean Patou	
F Blends		Pavi Elle by Avon	
Single F	H	Tea Rose Extrait by Perfumers Workshop	
Single F		Night Blooming Jasmine by Jovan	
AMB	H	Arpège by Lanvin	
AMB		Nikki de Saint Phalle by Cardin	
AWF	H	Nocturnes by Caron	
AWF		Paloma Picasso by Picasso	
Chypre	H	Miss Dior by Dior	
Chypre		Geminesse by Max Factor	
Orientals	H	Bal à Versailles by Jean Desprez	
Orientals	H	Youth Dew by Estée Lauder	

PERFUME SUGGESTIONS FOR ROMANCE AND SEDUCTION

Category		Name	Appropriate Products
G	H	Private Collection by Estée Lauder	Eau de Toilette Category
G	H	Givenchy III by Givenchy	Eau de Perfume Category
FFF	H	Armani by Giorgio Armani	Powders
FFF		Scoundrel by Revlon	Lotions
Sweet F		Echo by Borghese	Cremes
Sweet F		Jontue by Revlon	Bath Oils
Light and Soft F		Chanel No. 22 by Chanel	Bath Oil Sprays
Light and Soft F		Fleurs de Fleurs by Nina Ricci	(Be sure to carry a purse spray!)
F Blends	H	White Shoulders by Evyan	
F Blends		Giorgio by Giorgio	
Single F		Misty Tea Rose by Jovan	
Single F		Gardenia by Houbigant	
AMB	H	Ombre Rose by Jean Charles Brousseau	
AMB	H	Chanel No. 5 by Chanel	
AWF	H	Je Reviens by Worth	
AWF		Volcan d'Amour by Diane Von Furstenberg	
Chypre	H	Intimate by Revlon	
Chypre		Le Jardin by Max Factor	
Oriental	H	Opium by Yves Saint Laurent	
Oriental		Shalimar by Guerlain	
Animalic	H	Tigress Musk by Fabergé	
Animalic		Musk by Jovan	

I have combined perfume suggestions for all kinds of romantic evenings. From simple strolls to boudoir seductions, the fragrances listed encompass all moods. But remember, the closer you are to him the lighter your fragrance should be. Strong, H-marked fragrances can be a turn-off, so use the lightest concentrations —eau de colognes, splashes, powders, lotions—and layer them if you feel it necessary. Let your fragrance whisper your romantic mood.

Fragrance Sense has a language all its own. It speaks without words and becomes representative of what we feel about ourselves and what we tell others about ourselves. The messages we convey need to tally with our personality and lifestyle.

Developing your own Fragrance Sense is not such a difficult or expensive undertaking after all! Let's summarize the Four Steps:

* Learn the perfume pitfalls and the rules for buying perfume.
* Take the scientific perfume test to discover if you throw off scents, and study the product guide to see which fragrance concentrations become your best buys.
* Study and understand the eight basic fragrance categories, follow the perfume trail to discover your fragrance category, and use the name-brand fragrance guide to start yourself thinking about other fragrances in your category.
* Start building your fragrance wardrobe based on your lifestyle.

By following these steps, you will have developed fragrance habits that will serve you well the rest of your life. Use the knowledge you have gained to express your own personal Fragrance Sense!

PART III

You And Your Perfume

CHAPTER · 8

How Perfume Can Work For You

A ttracting the opposite sex isn't the sole function of fragrance, though it is the one stressed most by everyone who manufactures perfume. But fragrance in fact can help you when you are totally alone. If I were locked in a room alone for a week, I would still want to wear perfume because I think perfume is the ideal tool to make you feel better about yourself when you are by yourself.

PERFUME THERAPY

I first discovered fragrance when I was seven, as I said in the introduction. Using it made me feel fantastic, so every time I felt low about school or suffered from growing pains or the torments of teenage love, I would sneak into my mother's bedroom and sniff her bottle of Chanel No. 5. I would immediately feel better, recapturing the thrill of my first scent. You could call what I did perfume therapy. I have used the same technique through various stages of my life to make me feel good about myself and life in general.

Have you ever noticed the happy reaction you have to a certain smell that reminds you of something pleasant? For example, if I smell ham, peas, or biscuits I instantly feel as if I were four years old again and back in my Granny's kitchen in Florida, where she would bake my favorite foods every time I visited her. Old Spice shave lotion brings back memories of early Saturday mornings when I would watch my father shave and then splash on his fragrance. The smell of the lotion made such an impression on me that, for a while, I believed *all* men smelled of Old Spice. Chanel No. 5 makes me feel twelve, awkward and shy because the first time I went with a boy to an early-evening movie, my mother applied a little of her Chanel No. 5 behind my ears. Ever since then, no matter how many sophisticated Chanel commercials I see, the fragrance still reminds me of being awkward, shy, and twelve years old.

When I was in my late teens, I went to London for Christmas. I loved London: the pageantry, the elegance, and the peaceful, quaint quality of the city. Each night I would amble back to my hotel after a day of sightseeing and relax in a hot bath of Love's Lemon. My London days were only complete after I sank back and luxuriated in the refreshing lemony-smelling bathtub. Now, each time I take a whiff of Love's Lemon, I am instantly transported to Christmastime, at seventeen, in England.

Perfume therapy is a way to recapture moods and emotions and trigger memories. Annette Green, the executive director of the Fragrance Foundation, explains perfume therapy in more scientific terms: "Smell messages go directly to the brain. The area of the brain that receives these messages is the same area that receives and interprets our emotional and artistic responses *and* what we remember." The sense of smell triggers off memory even more strongly than sight does. The nose receptors lead directly to the limbic system, which controls the brain's mental images.

The British Red Cross has used fragrances for several generations as a way of warding off depression and minimizing stress. Doctors there discovered that when patients were given perfume therapy, they healed faster, didn't have relapses, and fitted back into society and a social life more easily than those patients who hadn't received perfume therapy.

A recent scientific study conducted by PFW, Incorporated re-emphasized the powerful association between the sense of smell and the brain's message center. Volunteers were asked to paint identical walls with two almost identical cans of paint. There was one small difference between the two cans of paint: one can was laced with a minute and undetectable amount of perfume; the other was unscented. After the walls were painted, the volunteers were asked to compare the quality of the two different paints. An overwhelmingly conclusive 100 percent of all volunteers claimed that the scented paint was smoother, covered the wall better, was easier to work with, and had a better color than the unscented paint. Not one of the volunteers realized that the paint they had picked was scented, and therefore far more attractive to them.

I have used perfume therapy not only to make me feel good about myself and improve my memory, but to keep my sense of smell functioning at peak performance. Give it a try. I promise you'll be able to smell the results and improve your sense of smell almost immediately.

All in all, fragrance has the power to alter and influence your feelings about everything, especially yourself. And ultimately, the way you feel about yourself will spill over to the way in which other people feel about you.

FRAGRANCE LANGUAGE

Perfume can create a tremendous aura of confidence in a woman, making her sparkle and dazzle everyone around her. We all experience moments of butterflies in the stomach, apprehension, tension, and just plain "nerves." I've found that spraying on a little bit of my favorite perfume before that stomach-lurching meeting or that heart-pounding first date can make me feel confident and able to take on anything. I am sure the same technique will work for you as well.

I always think of fragrance as a type of body language that subtly signals to the world who exactly we are and how we want to be treated. Of course that works in reverse as well. You can sometimes judge other people by the kind of scent they wear. For example: active women tend to select green, or sportlike, fragrances (Aliage by Estée Lauder, Cristalle by Chanel). Extremely feminine or romantic women often choose floral fragrances (Fleurs de Fleurs by Nina Ricci, Chloé by Karl Lagerfeld). Avant-garde or modern women seem to prefer the aldehydic modern blends of fragrance (Nikki de Saint Phalle by Pierre Cardin, Paloma Picasso by Paloma Picasso). I think choice of a fragrance sums up a woman so strongly that I pride myself, in a way, on finding out a great deal about a woman just on the strength of her perfume choice. I've become so practiced at doing it that I sometimes even refer to myself as a scent detective.

In other words, perfume really can project an image. In Chapter 7 you learned about how different perfume styles work best to enhance your image in different circumstances. Perfume has a language of its own that can help to define the self that we present to others. This is yet another of the many ways perfume can work for you.

CHAPTER · 9

Caring For Your Perfume

Contrary to popular belief, a fragrance doesn't last forever. The chemical properties of a fragrance change with time so that the scent is marred. A friend of mine still owns a bottle of Arpège that her husband brought her from Paris twenty years ago. It stands on her bathroom shelf, a sentimental souvenir that she will never be able to use. The Arpège has turned green, as aging perfume often does, and has lost its original luster.

* *Store* your perfume in a cool, dark place—in a closet or in the fridge if you like to apply it very cold. If a fragrance is exposed to heat, sunlight, air, or moisture, it will decay. When a fragrance begins to darken in color, you can be pretty sure that its scent has "gone off." Odors are created by the evaporation of moisture molecules. If you leave the cap off a bottle of

fragrance, it will eventually evaporate. A 2 ounce spray bottle of eau de toilette used daily will last three months. A 1 ounce bottle of perfume concentrate used once a week will last five to six months. (If you use it up faster, you're applying it too often. Slow down!) However, if you don't touch it at all, it will start to change after a year, even if you have never opened the bottle.

* On the subject of perfume *bottles* and perfume preservation: if you use a perfume decanter to hold your perfume, make sure that you wash it thoroughly in a solution composed of 50 percent water and 50 percent alcohol before you put a different scent into it. The alcohol is necessary to rinse away the residue of the scent you originally had in the bottle, and if you fail to use it, you will ruin your new scent.

* You can also destroy perfume by using glass stoppers with which to apply it. Glass is neutral and does not react with a perfume *inside* a glass bottle. The stopper is not for applying perfume but to protect it. If you rub a glass stopper onto your skin and then put it back into the bottle, you will transfer whatever is on your skin directly into the perfume, which will cause it to age faster. So apply the perfume with your fingers.

* If you take your perfume with you when you *travel*, don't transfer it into a plastic bottle because the alcohol and oil in perfume can eat plastic. Perfume is best transported in a leakproof glass bottle or jar. Use it on journeys by dabbing towelettes with your favorite scent. Or use purse perfume atomizers—but remember that these can also leak. I usually seal my atomizer in a plastic bag before putting it into my purse. Even though I may be wild about a fragrance, I don't want it to spill in my purse so that I am forced to smell it for the next ten years every time I open the bag. Perfume may initially appear to be a very

romantic subject, but I believe you have to approach it very practically!

* One fragrance area I haven't covered is mixing your own perfume. I'm afraid part of the reason I haven't mentioned it is that I am really not very keen on you mixing your own perfume. I think it is an expensive and complicated business, at the end of which you still won't have created anything that compares with fragrances you can buy. Mixing your own perfume is expensive because so many oils need to be mixed together to get a blend of scent. There are approximately 50 to 175 different oils in each perfume already on the market. If you try to match this, you will spend a fortune and may not even like the results. But if you *do* want a custom-made fragrance, I suggest you go to a custom-blended perfume shop because they have the expertise to blend the scent you want.

CHAPTER · 10

The Future Of Perfume

The technological advancements of the eighties have changed perfume along with computers, video machines, and so on. You could almost apply the phrase "state of the art" to perfume as well. I've already mentioned the latest innovations in the manufacturing of perfumes, including pheromones and the linear perfume, but a few very futuristic perfume developments are also currently being pioneered.

Jeffrey Miles, who is Director of Creative Services at PFW Division of Hercules, Inc., a major supplier of fragrances and flavors, says that researchers are starting to use fragrance as a way of helping people lose weight. "There is both a positive and a negative approach," he says. "The positive approach is to help people fight depression and anxiety, which are the biggest causes of overeating, by getting overweight people to smell their favorite fragrances as a way of warding off negative emotions." The other scent-induced weight-loss technique involves putting people off eating by making them smell an odor they hate.

Blind people are also being helped by fragrance in the 1980s. Companies are currently manufacturing wax containing micro-encapsulated fragrances. When the wax is used on a floor, a fragrance is released the moment anyone walks on it. That way, a blind person can know he is in the kitchen (for example) when the kitchen wax fragrance released has a bacon-and-egg smell attached to it. The same can be done with the den so that the den floor wax releases a cigar-scented odor if someone walks on it.

Another futuristic trend that may soon become popular is scented TV commercials. That's right! We will actually be able to *smell* television commercials in the future. You'll soon be able to watch a TV commercial for food, twirl a dial on the television set, and *smell* the aroma of the food being advertised.

In fact, perfume isn't just going to be sold in the future, it is also destined to be employed to sell. It is projected that clothing and department stores will soon be using fragrance psychology in various departments without your being aware of it as a way of subliminally enticing you to purchase products. So watch out—in a few years' time we all may *literally* be led by the nose!

I think I would like to try a new alarm clock invented by Dr. James Kavoussi and Louise D. Hartford. The scent-awake clock wakens you by spraying your favorite perfume into the air.

Aside from the more esoteric fragrance developments, I really am convinced that our general everyday concept of fragrance has altered radically in the past few years. When I was growing up, perfume was something you sparingly sprayed on before an ultra-important occasion. But it just isn't that way anymore. Colognes and perfumes are now part of every woman's daily grooming routine, something that to many women is as important as brushing their teeth and putting on makeup.

I've tried to pass on my passion for perfume and to inform, entertain, and enlighten you on as many aspects of perfume as possible. I hope I've succeeded. I've also tried to give you practical perfume facts. But sometimes you can't get away from another fact: perfume is a glamorous, romantic essence. No matter how down-to-earth one tries to be, sometimes the seductive potency of perfume just has a way of taking over and transforming the facts into something magical.

Perfume Glossary

Accord:
A blending of oils and extracts, aroma chemicals, and isolates. The chypre fragrance category is an accord.

Aroma chemicals:
Chemicals with specific aromas used in place of natural essences.

Bottom note:
Called "dry-down" by perfumers. It is the reaction of the residual elements in the fragrance to your skin. After a perfume made in the traditional three-layer process has been on your skin for an hour, the dry-down stage is reached. Amber and spice are found in the bottom note.

Concentration:
This is the percentage of essential oils in a fragrance. The higher the concentration, the longer-lasting the fragrance.

Di-propylene glycol:
An ingredient used in perfume, which is similar to graded mineral oil.

Distillation:
The process of bringing water and flower petals to the boiling point and collecting the desired essence as the steam condenses.

Essential oils:
The extracts from flowers, herbs, fruits, woods, grasses, beans, etc., that produce the actual fragrance itself.

Evaporation scale:
The measurement perfumers use to record evaporation times for fragrance produced by the three-layer method.

Fixatives:
The ingredient in a perfume that holds or "fixes" the fragrance.

Isolates:
Fractional extracts of fragrances used to build a perfume.

Isopropyl myristate:
Another oil base used in perfume, similar to d-propylene glycol.

Linear construction:
A new method of fragrance construction in which the top, middle, and bottom notes of the fragrance contain similar scents; Jean Naté, for example, smells lemony throughout the dry-down.

Maceration:
The soaking of successive batches of fresh flower petals in huge vats of suet or lard, from which the fragrant oily essence is then extracted.

Middle note:
The second layer of evaporation in a fragrance produced by the three-layer method, which lasts about an hour on the skin. Floral notes are often middle notes.

Natural essence:
Fragrance derived from flowers, stems, or animals.

Nose:
One of an elite group of expert perfumers with an acute sense of smell that allows him or her to identify many of the component scents (there are some three thousand available) in blended fragrances.

Note:
The term used in the perfume industry to refer to one of the various elements in a fragrance. It can be a type of scent or a layer of evaporation.

Perfumery:
The art of combining hundreds of different scents to construct a fragrance.

S.D. alcohol:
This is graded alcohol available (with a government permit) to create perfume.

Steam distillation:
A method of distilling fragrance from a flower.

Synthetic essence:
A combination of chemicals used to create a scent. The scent may duplicate a natural scent, or it may be a creation of fantasy—the perfumer's idea of the fragrance of a summer breeze or the sea before a storm, for example.

Three-layer method:
The traditional process of constructing a fragrance in which the top note may contain green scents, the middle note floral scents, and the bottom note amber scents. The perception of the fragrance changes as top note evaporates to middle and bottom note.

Top note:
The first smell you perceive from a fragrance created by the three-layer method. It lasts no longer than a few minutes on the skin. Fruity and green notes are often found in the top note of a fragrance.

Index